Ligand Efficiency Indices for Drug Discovery

Ligand Efficiency Indices for Drug Discovery

Towards an Atlas-Guided Paradigm

Celerino Abad-Zapatero

ELSEVIER

AMSTERDAM • BOSTON • HEIDELBERG • LONDON
NEW YORK • OXFORD • PARIS • SAN DIEGO
SAN FRANCISCO • SINGAPORE • SYDNEY • TOKYO
Academic Press is an imprint of Elsevier

Academic Press is an imprint of Elsevier
The Boulevard, Langford Lane, Kidlington, Oxford, OX5 1GB, UK
225 Wyman Street, Waltham, MA 02451, USA

First published 2013

British Library Cataloguing-in-Publication Data
A catalogue record for this book is available from the British Library

Library of Congress Cataloging-in-Publication Data
A catalog record for this book is available from the Library of Congress

ISBN: 978-0-12-404635-1

For information on all Academic Press publications
visit our website at elsevierdirect.com

This book has been manufactured using Print On Demand technology.

Working together to grow
libraries in developing countries

www.elsevier.com | www.bookaid.org | www.sabre.org

ELSEVIER   BOOK AID
            International   Sabre Foundation

Printed and bound by CPI Group (UK) Ltd, Croydon, CR0 4YY
Transferred to digital print 2013

To the future generations of drug discoverers:

May they find the correct variables and the most efficient courses.

CONTENTS

ACKNOWLEDGMENTS

Purposely, I have mentioned people who have contributed to the development of these ideas throughout the text for two reasons: (i) to provide a personal narrative making the content less arid; and (ii) to put their contributions in the appropriate context in time and space.

I would like to add a brief note of appreciation to them and to other persons that have helped to make AtlasCBS a reality. The initial ideas related to ligand efficiency grew out of discussions with Dr. Jim Metz at Abbott Laboratories. John Wass, neighbor and colleague, has always been my constant advisor in statistical matters. Dani Blasi and Jordi Quintana at the Platform of Drug Discovery in the Parc Cientific Barcelona used these concepts during my extended visit and we had insightful discussions. The initiative to support the creation of the AtlasCBS server came from Dr. A. Morreale (CBMSO, Madrid) and Prof. F. Gago (University of Alcalá de Henares). Alvaro Cortés-Cabrera did all the programing and transformed a "dream" into a reality. They also read the manuscript as it was unfolding and provided valuable comments. Dr. John Overington and Dr. Mark Davies were instrumental in installing it at the European Informatics Institute (EBI, Cambridge, UK) and Dr. Peter Rose orchestrated the link to the PDB. Dr. Ana Patricia Bento (EBI) learned the ligand efficiency concepts rapidly and used them to map the active peptides in the ChEMBL database. I'll treasure the hospitality and camaraderie of the entire ChEMBL group during my stay within their ranks. I appreciate the support of the Center for Pharmaceutical Biotechnology at UIC and in particular Prof. Michael Johnson. The encouragement of Profs. L. Tabernero and J. Bella at the University of Manchester, UK is also acknowledged.

Finally, the persons directly involved in bringing this modest opus to existence. Kristine Jones, Senior Acquisitions Editor for Elsevier, supported the project enthusiastically from the very beginning and Andy Albrecht was always ready to answer technical and editorial questions along the way. Ms. Lisa Jones and her colleagues had to collate the multitude of images and convert the manuscript into a "mini-atlas" with instructions. I appreciate the assistance of Ms. Tina Mistry with the figures and Ms. Elizabeth Woods in proofreading sections of the manuscript. The personal support of family, friends, and colleagues while the manuscript was being written is greatly appreciated.

THE CURRENT STATUS OF DRUG DISCOVERY

The systematic search for chemical entities with therapeutic effect began in the late 1800s and was connected to the development of organic chemistry and organic synthesis in Germany. The milestone of these efforts was the synthesis, testing, and effective use of Salvarsan, the first successful treatment for syphilis in 1911.[1] The dramatic advances in biochemistry, biology, and genetics in the first half of the twentieth century provided a more sound basis to the relationship between the biochemical processes taking place in the living organisms and the active chemical entities. Later, the discovery, successful use, and commercialization of penicillin and other antibiotics launched the birth and development of the pharmaceutical industry as we know it today.

The advances in molecular biology and structural biology, in particular the unveiling of the first protein structures (myoglobin and hemoglobin) in the early 1960s, opened the door to a new way to approach drug discovery. Structural biologists gained more confidence as the number of protein structures and protein–ligand complexes increased: structural biology of drug-discovery targets and ligands could contribute to design drugs more expeditiously or even "rationally." There was a sense of optimism. On the other side, the completion of the human genome and of the genomes of pathogens (i.e., *Mycobacterium tuberculosis*) appeared to open a new era in drug discovery. However, this initial excitement has been dampened. The inherent complexity of biological systems is still a major hurdle hampering our progress.

One can point to the article published in 1976 by Beddel and colleagues[2] as the first structure-based design of compounds directed toward a site in human hemoglobin based on its 3-D structure. The first protein-docking example was published a few years later (1982) by Kuntz and his colleagues.[3] These two methodological innovations, enhanced by the exceptional developments in computer hardware and software, are today part of the mainstream of drug discovery in both the pharmaceutical industry and the academia. Computer-assisted drug design (in all its facets) is affirmed as a critical tool for the future development of drug discovery, based on the extensive knowledge of protein structures that have been seen in the last 50 years. Experimentally, fragment-based approaches are becoming more widespread but their effectiveness is still not fully realized.[4,5]

The initial enthusiasm on the successful application of computational approaches to *de novo* drug discovery, based on the calculation of binding energies between the target and the ligand, dampened in the late 1990s upon the realization that the problem was more difficult than anticipated. Technical problems such as scoring functions, force fields, entropy contributions, calculation of binding energies in the solvated state are still unresolved in a definitive manner.[5] In addition, even if a compound can be optimized for binding to its cognate target, additional issues need to be solved in relation to its absorption, distribution, metabolism, and toxicity (ADMET) properties. Although approaches are currently in place to address these problems up front, subtleties always appear in the last stages of preclinical development where the outcome is not always predictable. Undoubtedly, although we have made progress in finding, designing, synthesizing, and developing drugs, our tools and methods are far from optimal. There is much that needs to be done, conceptually and technically to make drug discovery more effective.

A novel approach to guide drug discovery is presented in the pages that follow. It is based on the notion that a "change of variables" is necessary. Affinity between the chemical entities (ligands) and the biological macromolecules (targets) has always been a key parameter in drug discovery. However, ligands need to have certain properties to be successful drugs. An algebraic framework is proposed that combines the affinity with the physicochemical properties of the ligands, as a set of novel variables to map chemicobiological space and to optimize the drug-discovery process.

REFERENCES

1. Sepkowitz KA. One hundred years of Salvarsan. The New England journal of medicine. 2011; **365**(4): 291–3.

2. Beddell CR, Goodford PJ, Norrington FE, Wilkinson S, Wootton R. Compounds designed to fit a site of known structure in human haemoglobin. British journal of pharmacology. 1976; **57**(2): 201–9.

3. Blaney J. A very short history of structure-based design: how did we get here and where do we need to go?. Journal of computer-aided molecular design. 2012; **26**(1): 13–4.

4. Hajduk PJ, Greer J. A decade of fragment-based drug design: strategic advances and lessons learned. Nature reviews drug discovery. 2007; **6**(3): 211–9.

5. Jhoti H, Leach AR. Structure-based drug discovery. Dordrecht: Springer; 2007.

INTRODUCTION

Navigating rough and unchartered waters is extremely dangerous. In the eighteenth century, England (*Longitude Act enacted in 1714*) and other seafaring nations of the world (Spain, the Netherlands among others) posted a prize equivalent to a king's ransom in gold to the person/institution that could solve the most pressing technical problem of the world at the time: finding an accurate, "practicable and effective" method to determine the longitude at sea. Just few years earlier (1707), the British Navy had lost four homebound warships near the Scilly Isles of the expedition of Admiral Sir Clowdisley Shovel because, when in the middle of the sea and not knowing where the fleet was, aboard his *H.M.S. Association* he made the biggest blunder of judgment of his long career with the British Navy. He ignored the advice of a humble seaman as to the fleet's position. He had him hanged for mutiny on the spot and triggered the catastrophic event that took nearly 2000 lives[1,2].

After 40 years of knocking at doors and trying to convince the British establishment, the humble clockmaker John D. Harrison solved the problem with his successive generations of chronometers. These devices permitted an accurate charting of the globe. Most likely, it was the superb charts of the British Admiralty that gave the British Navy its superb mastery of the seas. All derived from successive mappings of the world's seas by several meticulous expeditions using Harrison's chronometers, among them the *H.M.S. Beagle* of Darwin fame.

Putting aside differences, navigating the rough waters of drug discovery even in the twenty-first century is also dangerous. Decisions are made at the management and laboratory level that are not based on reliable variables that permit a safe arrival to the venturous port of a safe yet effective drug in a reasonable amount time. Affinity variables toward the biological targets (K_i, IC_{50}, and others) as well as the physicochemical parameters of the chemical entities (PSA, MW, C log P, "Rule of Five" criteria, and others) are kept, maintained and updated in massive databases in attempts to "map" the territory. However, numbers, *per se*, do not make good maps. It is important to find the correct set of variables that provide "insight" into the processes and select the most suitable combinations to make clear charts.

Any high school chemistry student can grasp the power of the concept of density (mass/volume) as being superior to the two separate variables. Then, chemical processes and phenomena can be described numerically and mathematically as being dependent on *density* not on mass *and* volume. Through this work, I wish to suggest that ligand efficiency indices (LEIs) can play a similar role and that in the not too distant future, strategies can be proposed and effective trajectories may be drawn in chemicobiological space (CBS) that will make the drug-discovery process "safer," more cost-effective and efficient, and not just a "numbers game" anymore.

I am confident that if we could find the correct set of LEIs to map, describe, and guide the drug-discovery process, we would be a long way toward making it more effective, expedient, and economical. A reliable charting of the extent and complexity of CBS would represent a tremendous aid in charting our expeditions through the rough waters of drug discovery.

A few years ago, an apparently simple "seed" was introduced: ligand efficiency. This is the notion of exploring how the potency of a compound relates to its size and how it can be used to compare different compounds as fragments or as leads in the drug-discovery process. The concept has grown in different laboratories in academia and industry to suit the needs of the drug-discovery practitioners, each one using their own definitions and applying them to make decisions. Those definitions would be summarized and compared when possible. As we know, it is difficult to compare apples and oranges and the relative merits of the different definitions of ligand efficiency are still under exploration.

However, the main emphasis of this opus would be to present a set of unified efficiency variables that permit an effective mapping of CBS. The descriptive properties of this representation will be presented, illustrated, and discussed attempting to show its power and its current limitations. The applications of this representation to various areas of drug discovery will also be illustrated and an attempt will be made to draw conclusions and to present some perspectives for the future. This oeuvre will put forward the notion that LEIs can be used as a new set of coordinates to navigate through the tortuous waters of drug discovery using an atlas-like representation of the CBS. In this work, only the visual and intuitive aspects of the framework will be presented. However, I surmise that the subjacent algebraic and spatial (multidimensional) framework could provide also a scaffold for a more robust set of tools to guide drug discovery in the near future.

REFERENCE

1. Sobel D. Longitude. New York: Walker Publishing Company, Inc.; 1995.
2. Wilford JN. The Mapmakers. New York: Vintage Books. A Division of Random House; 1982.

PART I

The Elements: Data, Variables, Concept, and the Server

The vast amount of data relating the affinity of small molecules (ligands) to the biological molecules to which they have any activity (targets) is now stored in several public and private databases of considerable complexity and magnitude (Structure–Activity or SAR Databases: BindingDB, PDBBind, ChEMBL, PubChem, WOMBAT, and others). These data comprise the currently known chemicobiological space (CBS). Although comprehensive in scope and extent, the databases themselves do not provide dramatic insights as to the basic principles that govern the interaction between ligands and targets, nor do they provide a graphic road map to guide the drug-discovery process.

Ligand efficiency indices (LEIs) represent a novel way to look into the affinity of the ligands toward their targets by taking into consideration the affinity, size, polarity, and other physicochemical properties using combined variables. The various formulations are presented and discussed and a combined framework is introduced that permits graphing the content of the SAR databases in an atlas-like representation. A web tool (AtlasCBS) is described that allows the user to map CBS online and summarizes the content of those databases in efficiency planes. This representation is akin to an atlas of CBS that could guide drug-discovery efforts in the future.

CHAPTER 1

Structure–Activity Databases for Medicinal Chemistry

On October 30, 2011, the community of macromolecular crystallographers and structural biologist gathered at Cold Spring Harbor Laboratory to celebrate the 40th anniversary of the origins of the Protein Data Bank (PDB), the most notable of the three-dimensional structural repositories of macromolecules of biological interest. It had been 40 years since the concept of a public access collection of protein structures took hold in the incipient community for protein crystallographers. The spectacular growth and development of this global resource has been described by one of the pioneers of the PDB[1]. The idea of a common depository (a database) of crystallographic results relevant to the biological and chemical community was a visionary concept. These were the days well before the Internet was invented and a "data set," a protein structure, was contained in a box of about one thousand IBM computer cards (80 columns across), one card per atom, with the three Cartesian coordinates for each atom.

Since then, many databases have developed within the scientific community at large and the advent of the Internet has given these public resources tremendous flexibility of content, access, and interrelation.

For the purpose of drug discovery, three kinds of databases are of special relevance: chemical, biological, and the most recent ones that relate both. Chemical databases typically store small molecule structures of both drug and nondrugs, usually linked to the commercially available chemical compounds and they represent a tremendous resource for virtual or experimental screening (e.g., ZINC)[2]. The content of biological databases is predominantly gene nucleotide sequences and/or protein amino acid sequences as they relate to their biological function. Most important among them is the originally named SwissProt[3], which now contains over half a million protein sequences and has evolved into a formidable bioinformatics resource to become the Swiss Institute for Bioinformatics (SIB) (www.isb-sib.ch). The increasing accumulation of structure−activity data, originating from various *in vitro* and *in vivo* assays of a myriad of small molecules, targeting a growing number of biological macromolecules from an increasing number of biological targets, has resulted in databases that are referred to as large-scale SAR databases. These databases typically relate a document (D, journal or patent) reporting the data or the result(s) from one or various assays (A) of a certain compound (C, small molecule), on a certain protein target (P) with a quantitative result (R), given in concrete units[4,5]. Various SAR databases obtain (or extract) these data and organize the content in different ways to facilitate access to the information. This chapter provides an overview of the most prominent databases (private and public) of interest for drug discovery; it is by no means meant to be exhaustive. It focuses on the databases most relevant to the drug-discovery community and the ones that have been used for the development of the ideas and concepts presented here.

The most veteran public chemical databases, Chemical Entities of Biological Interest (ChEBI) and PubChem, made their appearance in 2004 and have grown and developed considerably since that time. ChEBI is a database of endogenous and synthetic bioactive compounds that features a chemical ontology classification of the chemical entities[6,7]. PubChem is a massive open repository of experimental data that is organized in three distinct databases: PubChem Substance, PubChem Compound, and PubChem BioAssay. Only the PubChem Compound branch contains the strictly chemical information with links to the chemical substances and bioassay data. It also provides links to the literature citations via PubMed. From these initial efforts, a large variety of databases of interest to the drug-discovery community have matured. A brief description of the most relevant (public and private) SAR databases follows. Their order reflects the historical path that the different databases have played in the development of the ideas presented in this book.

1.1 PDB

As a protein crystallographer, my early contact with databases originates from the existence of the PDB. The PDB is a unique database because since its inception, the focus has been to make the three-dimensional structures of proteins and macromolecules available to the community of structural biologists at large. In the early years of protein crystallography, the solution of each new structure was the result of several years of hard labor by a committed group of researchers and the concept of making those results publicly available was not an obvious thought. Generally speaking, the growth of the PDB has been exponential through the years with the number of entries doubling approximately every 2.75 years[8]. However, there was a slow down of the deposition rates in the mid-1980s related to the relevance and importance of macromolecular structures for drug discovery. Details of the history of the PDB can be found in the historical perspective by Berman[1] and a more detailed analysis of the deposition rates can be found in a brief essay commemorating the 40th anniversary of its birth[8].

Currently, the PDB is the central repository of macromolecular structures including proteins, nucleic acids, and virus structures, as well as a rather large number (>6,000) of protein–ligand complexes, determined by experimental methods. Ligands are any small molecule that binds to the protein/macromolecule in the crystalline environment. Although the vast majority of the structures have been determined by X-ray crystallographic methods, entries now also include structures determined by alternative physicochemical techniques such as nuclear magnetic resonance (NMR) and high resolution electron microscopy. It should be noted that although at the beginning the PDB was physically located at Brookhaven National Laboratory (Long Island, NY), since 2003, the PDB is a global organization called the Worldwide PDB[9] (www.wwpdb.org). It has four centers distributed on three continents: Research Collaborative for Structural Bioinformatics (RCSB) and the BioMagResBank (BMRB) in the USA, the Protein

Table 1.1 PDB Current Holdings

Method	Protein[a]	Nucleic Acid	Protein/NA	Other	Total
X-ray[b]	68,887	1,389	3,482	3	73,761
NMR[c]	8,362	1,003	189	7	9,561
Electron microscopy	306	22	120	0	448
Hybrid	44	3	2	1	50
Other	141	4	5	13	163
Total	77,740	2,421	3,798	24	83,983

[a]*Protein includes protein, viruses, and protein–ligand complexes. As of August 21, 2012, 5 p.m. PDT.*
[b]*63,188 structures in the PDB have a structure factor file.*
[c]*6,868 structures have an NMR restraint file. 628 structures have a chemical shifts file.*

Data Bank Japan (PDBj) and the Protein Data Bank in Europe (PDBe; PDBe.org). The four wwPDB partners accept, process, and curate new structures (further details can be found at Velankar et al.[10]). A quick consultation of one of the PDB websites (www.rcsb.org) reveals that at the time of writing, the number of entries is 83,983 (13,887 target–ligand complexes). Of those structures, 73,761 were determined by X-ray methods, 9,561 by NMR, and 448 using electron microscopy (Table 1.1).

Each entry is defined by an alphanumeric access code (i.e., 1A4G) that gives access to the atomic coordinates, crystallographic details of the structure determination including the structure factors if available, and a few key references to the published report of the structure. From these key items, the PDB resource branches off to provide a tremendous number of options (too many to list here) that help the user digest, expand, and relate the information on the protein (target and related entries) and the available ligands (if any) including salts, ions, precipitants, and other pertaining information. The PDB is a key hub of information consulted routinely by other databases that relate proteins (targets) to the ligands interacting with them.

1.2 PDBBind, MOAD

The expanding knowledge of biomacromolecular structures (including complexes) and the advances in computational and *in silico* tools for drug discovery prompted the almost simultaneous development of two databases derived from PDB: PDBBind[11] and MOAD (Mother of all

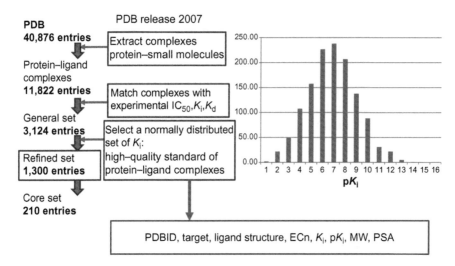

Figure 1.1 PDBBind 2007 extraction and content. Flow used in the extraction of the PDBBind (2007) set used in the ligand efficiency analysis presented in this work (see Chapters 2,5). The inset shows the distribution of K_i values of the set. Note the Gaussian distribution of the pK_i values ($pK_i = -log K_i$) and the broad range ($1 \geq pK_i \leq 14$). The data variables extracted for further analysis are enclosed in the red rectangle. Ligand structure was stored as a SMILES string; ECn refers to the Enzyme Commission number to identify the target. MW: Molecular Weight, PSA: Calculated Polar Surface Area.

Databases)[12]. These resources are based on the curated and extended content of the PDB with the specific purpose of aiding computational chemists to test, validate, and improve their software tools.

In both cases, the objective was to extract a subset of well-refined structures from the existing target–ligand complexes in the PDB and complement these data with the corresponding binding affinity data (K_d, K_i, and IC_{50}) from the primary references. A very important consideration in assembling the content of these databases was that the affinity data (most commonly K_i) would be randomly distributed over a wide range of affinity values (13 orders of magnitude). In the initial report of PDBBind[11], the database was built from 5,671 protein–ligand complexes extracted from 19,621 experimental structures existing in the PDB. Combining the structural results with the collected affinity data, the initial holdings of PDBBind were 1,359 complexes. The highest resolution structures (≤ 2.5 Å) were the refined set and the remaining 559 (≥ 2.5 Å) were also available as a secondary set. Figure 1.1 summarizes the corresponding selection and extraction process for PDBBind (2007), which was the database used during the early stages of application of the concepts of ligand Efficiency Indices (LEI) to explore the content of chemicobiological space (CBS). The most recent content, applications, and options of **PDBBind** can be found on their website (www.pdbbind.org). MOAD was built under the same principles but offers two significant content differences: the protein−ligand complexes are less redundant (only one protein per family is represented) and it includes protein−cofactor complexes[12] (http://www.BindingMOAD.org).

1.3 ChEMBL

My interest in the content of what would become ChEMBL began when I met John Overington at a drug-discovery forum in Edinburgh, Scotland, in 2006. On July 2008, the Wellcome Trust awarded a major grant to the European Bioinformatics Institute within the European Molecular Biology Laboratory (EMBL) at Hinxton, UK (EMBL-EBI) to fund the transfer of a rather extensive database from a publicly owned company (Galapagos NV) to the public domain. The data were incorporated into the EMBL-EBI which maintains and develops a major collection of open-access biomedical resources related to whole genome sequences. This action was at the center of a concerted effort to translate information from the human genome to practical outcomes, particularly in the domain of drug discovery. This was the origin of the ChEMBL database that constitutes a massive collection of fully annotated and curated information relating drug-like bioactive compounds to their biological targets. Like most of the other databases, the data are manually abstracted from primary scientific journals periodically and standardized. The most recent publication on the content of the database reports that it contains 5.4 million bioactivity measurements for more than 1 million compounds related to 5,200 protein targets[13]. The data can be accessed through their web-based interface, allowing convenient data downloads, and extensive web services within the resources of the EMBL-EBI laboratory (https://www.ebi.ac.uk/chembldb). To facilitate the navigation through the chemical−biological content, a series of "thematic portals" have been developed that direct users with individual interests, either by research topic or diseases and conditions. Thus, KinaseSARfari caters to the interests of the protein kinases community and GPCR SARfari focuses on the diseases and chemical entities connected to the G-protein coupled receptors. Of particular interest and related to "neglected diseases" is ChEMBL-NTD, which is a repository with open access to primary screening data for endemic tropical diseases of developing regions

(Africa, Asia, and the Americas, i.e., malaria). These data were donated by GlaxoSmithKline, Novartis, and St. Jude Children's Research Hospital to provide a freely accessible, permanent archive of high-quality screening data. These portals are engineered so that they can be easily adapted to other therapeutic areas of interest or new genes and gene families in the future[6].

1.4 WOMBAT

The WOrld of Molecular BioActivity (WOMBAT) database and its sister WOMBAT-PK are commercial databases that had their origins in a data-gathering project spearheaded by AstraZeneca R&D that focused primarily on results published in the *Journal of Medicinal Chemistry*. Even in 2006, the breadth and scope of the content of WOMBAT was impressive[14]. WOMBAT 2006.1 contained 154,236 entries (136,091 unique SMILES [Simplified Molecular Input Line Entry System] entries[15,16]), including 6,801 chemical series from more than 6,791 papers reporting in excess of 307,700 activities for 1,320 unique biological targets. Targets were clustered in four types: *receptors* (including GPCRs), *enzymes* (connected to their E.C. numbers), *ion channels*, and *proteins* (biological targets not included in the other categories). The data were gathered from four key medicinal chemistry publications: *Journal of Medicinal Chemistry*, *Bioorganic Medicinal Chemistry Letters*, *Bioorganic Medicinal Chemistry*, and *European Journal of Medicinal Chemistry*. All the data are organized under three different panels, each focused on a critical phase of drug discovery: bioactivity summary, target and biological information, and computed chemical properties. Of particular relevance for this discussion is that the third panel includes the calculated physicochemical properties of the ligands. In addition to the structure window, these properties are displayed in six windows that show values for molecular properties such as different number and types of atoms, Lipinski Rule of Five (Ro5)[17] parameters, Polar Surface Area (PSA), ALogPS, and the initial definition of ligand efficiency (LE)[18]. WOMBAT is an evolving database and additional content and features are updated regularly (http://www.sunsetmolecular.com). Compound structures, activities, organism(s) of the targets, and the assays performed with the resulting values are all available, combined with an extensive set of physico-chemical descriptors and molecular properties of the ligands (both experimental and calculated). The number of data items in WOMBAT is staggering. The 2012.1 version contains 331,872 entries (268,246 unique SMILES strings) representing 1,966 unique targets. The data are extracted from the available data in journals (15,320 papers) related to medicinal chemistry between 1975 and 2009. Most of the targets ($\sim$90%) listed in the entries that were obtained from source journals were linked unambiguously to SwissProt ID identifiers.

WOMBAT-PK shares some data with its sister database but contains additional unique data related to clinical pharmacokinetics (PK) and toxicology, including side effects; both are available in the MDL Isis/Base format. WOMBAT-PK-2010.1 contains 1,260 entries (1,260 unique SMILES) and includes more than 9,450 clinical PK measurements as well as extensive data on physicochemical properties, toxicity endpoints, and over 2,100 annotated drug-target bioactivity measurements. These data will be critical to establish a robust relationship between favorable PK properties and the concepts and definitions of LEI. Some of the results and applications presented later in this work (Chapter 5) used the data contained in WOMBAT (2007.1 release) kindly provided by Professor T. Oprea.

1.5 DrugBank

Although the information contained in this database was not used directly in illustrating the concepts and examples discussed later in this book, it is recommended that the reader consult this immense resource of information related to drugs and their targets.

DrugBank is a thoroughly annotated and vast database of information pertaining to drugs and biological targets. It contains extremely valuable information regarding the nomenclature (in different countries), ontology, chemical properties, structure, function and physiological action, pharmacology, PK, metabolism, and pharmacological properties of both small molecule and large (biologics) drugs. In addition, it contains well-organized and properly linked information on biological targets, mode-of-action, proteins, genes, and the organisms upon which the drug acts. It is indeed an integrated and comprehensive resource tool, indispensable for research relating drugs to any kind of "omics"[19]. All this information is readily available, organized, and connected via the individual "DrugCard" for each drug included in the database.

DrugBank first appeared in 2006 and is now widely used by academic and pharmaceutical researchers, clinicians, students, educators, and even the public at large. The latest release (DrugBank 3.0) has added more than 40 new data fields per drug (to total 148) and now includes drug-action pathways, drug transporter data, drug–drug interactions, metabolic degradation, side effects, and Absorption, Distribution, Metabolism, Excretion and Toxicity (ADMET) data. Other databases (ChEMBL, WOMBAT, BindingDB) also provide data on drug-related compounds, but DrugBank is currently the pinnacle of clinically related information on existing drugs. The latest version is linked to two critical databases of interest for this work (BindingDB, ChEMBL) and has been reciprocally linked to PubChem and other major chemical and biological resources. For reference, Table 1.2 lists a few of the limited (although relatively large) number of chemical entities available in DrugBank 3.0 that are approved for use as therapeutic agents in relation to the targets and other chemical entities.

An excellent way to navigate the labyrinth of drug-related questions within DrugBank is the "Browse" tab. This option allows users to search terms such as "drug clusters," "indications," or "drugs classes" and provides hyperlinks to >70 drug classes. The user can explore the content of the database from the perspective of: drug name, pharmacological class (i.e., antihypertensives), gene or target protein, pathways (~50), chemical classes (~240), and association (brief description of biological action).

Table 1.2 Relevant Numbers from DrugBank 3.0 related to Existing Drugs	
Number of Food and Drug Administration (FDA)-approved small molecule drugs	1,424
Number of biotech (biologics) drugs	132
Number of experimental drugs[a]	5,210
Total number of experimental and FDA drugs	6,816
Number of approved drug–drug targets (unique)	1,768
Number of all drug targets (unique)	4,326
Number of nutraceuticals drugs[b]	82
Number of experimental and FDA small molecule drugs	6,684

[a]Experimental drugs include: unapproved drugs, delisted drugs, illicit drugs, enzyme inhibitors, and potential toxins (>3,200 entries).
[b]Nutraceuticals or micronutrients such as vitamins, metabolites, and dietary supplements.
Adapted from Knox et al.[27]

1.6 BindingDB

The critical importance of this SAR database became apparent later on as I embarked on deeper analysis of the relation between the chemical and the biological worlds. BindingDB began at the University of Maryland in the late 1990s, and came on line at the end of 2000, probably the first publicly accessible affinity database. Officially, two papers in 2001[20,21] announced the birth of this database that currently complements and extends all of those previously mentioned. More technical details were reported a year later[22] establishing this database as a basic resource that we do not have space here to discuss in detail. However, it is a key component of the AtlasCBS server so more details will be introduced later in this work. There are many unique features and resources of this database that make it a critical tool for the concepts and applications that will be presented and discussed.

BindingDB is, first and foremost, a "binding database." It was designed as a database to hold as much "affinity data" as possible that were obtained by many different experimental techniques. Significantly, it contains data tables for ΔG, ΔH, $T\Delta S$, heat capacity, and equilibrium constant obtained from isothermal titration calorimentry (ITC) in addition to the more conventional K_i, IC_{50}, and K_d constants. BindingDB contains very valuable data related to enzyme inhibition because it is the most common assay related to enzyme activity and also because these results are of great pharmacological interest. The available data includes affinity measurements for compounds, series, and targets well beyond what is available from the scientific literature (including patents) and this is particularly important because, out of necessity, journals and patents have to limit the number of data tables in the publications. This treasure trove of additional data are obtained typically by direct deposition from the scientific teams involved in drug-discovery efforts both in academia and in private industry. It also offers chemical structure, substructure, and similarity searches for small molecules either alone or in complexes.

In addition to the compound binding data within BindingDB, another complementary component is the relation and search capabilities that it has regarding sequence homology searches in the biological or target domain. Currently, there are 1,717 protein—ligand crystal structures with affinity measurements for proteins with 100% sequence identity and if one allows a lower identity (85%), the number of structures increases

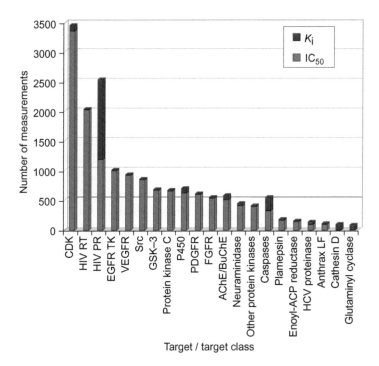

Figure 1.2 Number of entries in BindingDB for various targets (2006). A snapshot of the content of BindingDB (2007) in relation to the number of targets (or target class)[28]. A few targets (or classes) of high biomedical interest skew the data distribution (see Figs. 1.5A,B for more updated information but less detail on the specific classes). Reprinted with permission from Nucleic Acids Research, Copyright Oxford Journals.

to 4,937. These structures are properly annotated and linked to the corresponding chemical ligands facilitating further exploration of SAR relationships for a variety of targets of biological interest (Fig. 1.2).

Data are deposited into BindingDB by three channels: (i) literature extraction, deposition, and curation by on-site staff; (ii) deposition of new data at or near publication time by off-site researchers in a mode similar to the PDB; and (iii) high-throughput data uploads from computer-controlled instruments having software compatible with BindingDB.

In addition to all its internal resources of searching and downloading, BindingDB is thoroughly linked to the other key public affinity databases available to the community, namely ChEMBL and PubMed, as well as PDB. There are excellent user aids, documentation, and instructions on how to extract and download both the limited datasets and the entire content. Further information on the content of BindingDB, its relation to the other databases, and a general perspective is presented below.

1.7 PubChem

This database (http://www.pubchem.ncbi.nlm.nih.gov) originated through a US Government initiative started in 2004 by the National Institutes of Health (NIH) within the National Center for Biotechnology Information[23]. Originally, it focused on gathering activity data from the high throughput screening (HTS) programs supported by an NIH program called NIH's Molecular Libraries Roadmap Initiative, and later it added data from other sources. Currently, it includes BindingDB and ChEMBL, and thus its content also reflects data extracted from the scientific literature. The most up-to-date holdings are: 33 million distinct chemical entries and activity data from about 4,800 NIH assays[24]. Other details can be found at http://pubchem.ncbi.nlm.nih.gov/help. html#PubChem_Overview.

Conceptually, PubChem is organized into three different categories: compounds, substances, and bioassays. A specific chemical entity can be found in both the compound and the substance categories. Compound reflects a single standard representation, as opposed to substance that refers to the material used for the biological assay. Consequently, a given compound can exist in different substances as a component of the material that was tested.

The tools and resources for browsing, querying, and downloading data have been described elsewhere and it is not the purpose of this introduction to discuss the databases in great detail. Moreover, the content of PubChem has not been used extensively for the preparation of the examples described later in this book. Thus, recognizing the immense importance of this resource for the future of medicinal chemistry, we encourage the interested reader to visit and use this database.

1.8 PROPRIETARY DATABASES

An additional set of databases should be noted, not because of their open-access policy, but because they represent the private databases of the major pharmaceutical companies, and contain vast amounts of information. It is well known that the published data of successful and unsuccessful drug-discovery projects within the pharmaceutical laboratories is only a fraction of the available data. Typically, only one or a few of the compounds designed, synthesized, and tested during the long road to a successful candidate are released to the scientific literature in an attempt to rationalize a "path" from initial hit, through lead optimization and on to successful product in the clinical setting. These data include not only the *in vitro* testing but also the results of assays in cell and animal models. Proprietary databases also include extensive ADMET data and the extremely valuable data on the compounds or compound series that did not make it into the clinic for a variety of reasons. It is hoped that as the drug-discovery tools of the community at large improve, some of these datasets will be made available to researchers to expand datasets that could be used as valuable examples to develop sound and efficient strategies for rapid drug discovery. We should overcome the notion that drug discovery is "a numbers game" and achieve effective (preclinical and clinical) drug-discovery strategies and tools so that the process is expedited in time and resources to reduce costs and uncertainties to the patients and to society at large. As will be presented later, some of the Internet tools recently developed allow the uploading of proprietary data and their comparison with the sets available in public domain databases. Hopefully, the drug-discovery community (in academia and major pharmaceutical companies) will benefit from these new tools and concepts. In turn, these efforts will result in a more predictable outcome of any drug-discovery project.

1.9 PERSPECTIVE

While this brief section on databases was being written, an excellent perspective article was published in the *Journal of Medicinal Chemistry* by Gilson and collaborators[25] that certainly eclipses the content of this introduction. This is a timely review that puts the content, relevance, and critical importance of public domain SAR databases in perspective. In addition to the most relevant databases for the ensuing work presented here, the publication also introduces other databases (SuperTarget, ChemSpider among others) that are impossible to discuss within the limited scope of this work. I strongly encourage the reader to access this review and have it as a reference. Only the most significant trends extracted from this work will be summarized here to set the stage for the following discussion.

The number of unique small molecules and corresponding affinity data published annually has increased since 2008 (Fig. 1.3). The rate of increase in the publication of medicinal chemistry data has been estimated to be 50% per year (Fig. 1.4). As reflected by the content of

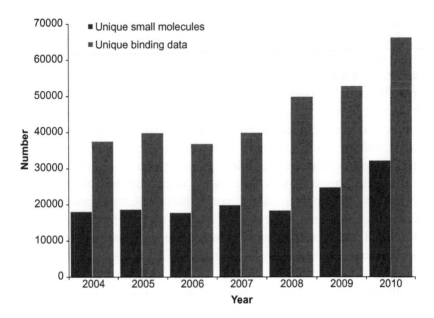

Figure 1.3 Publication trends of small molecule and related affinity data. Observed trends in the published unique data for small molecules and their associated activities by year. Reprinted with permission from Nicola et al.[25]. Copyright 2012 American Chemical Society.

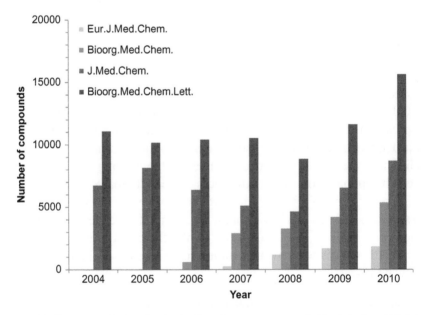

Figure 1.4 Publication trends of small molecule affinity data in various journals. Trends in the published data related to unique small molecules and their associated activities for the most prominent medicinal chemistry journals. The data are combined from BindingDB and ChEMBL. Reprinted with permission from Nicola et al.[25]. Copyright 2012 American Chemical Society.

BindingDB, the number of binding measurements per compound has an extremely skewed distribution with approximately 180,000 compounds with one affinity measurement and a long tail of a limited number of compounds for which hundreds of measurements are available (Fig. 1.5A). Alternatively, the distribution of the number of targets having a certain number of chemical entities for which affinities have been measured is also rather skewed (Fig. 1.5B) with a significant peak around 40 compounds target. The data on the number of measurements per compound and the number of protein targets having some compound for which affinity data are available is rather "punctuated" with a few hundred targets (~ 450) with two affinity measurements and a long tail corresponding to targets for which hundred (or even thousands) of compounds have been tested. This is a reflection of the specific focus of the drug-discovery community on "workable" (or druggable) targets and on the testing of known (and effective) chemical matter on a multitude of diverse targets.

There are two additional insights from the perspective by Gilson and colleagues[25]. One, the distribution of the molecular weight of the compounds from 2007 shows a dramatic increase in the number of compounds with low molecular weights (200–400 Da range) probably reflecting the increase in the application of fragment-based approaches to drug discovery. Second, leaving aside the comments made before about the large amount of confidential corporate data, the analysis documents that the published literature contains an equal amount of academic and corporate affinity data. Moreover, the amount of compound activity data in BindingDB, PubChem (BioAssay), and ChEMBL, including only measurements with a defined protein target, which originates from corporate laboratories, is larger (431,147, $\sim 60\%$) than the corresponding one from academia (303,321, $\sim 40\%$).

1.10 DISCUSSION

As impressive as the overview of the current databases linking chemical and biological data might seem, it is important to keep in mind a critical vision about their value, limitations and, on a more broad perspective, their significance. There is no doubt that the availability of these data in private and public databases is critical for further progress in understanding the interplay between chemical and biological space. The issues of redundancy and uniqueness between the content of the available

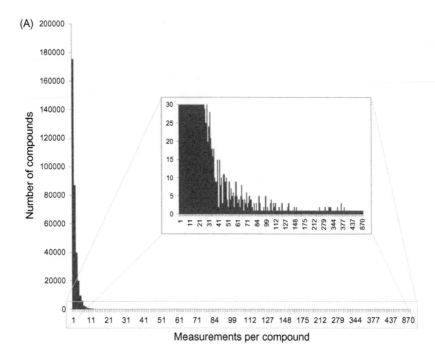

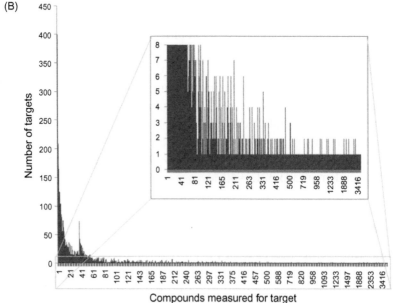

Figure 1.5 Number of binding measurements per compound in BindingDB. (A) Histogram of the number of binding measurements per compound in BindingDB. As of 2010, there were approximately 180,000 compounds with only one activity measurement. Inset shows the fine-grained details of the distribution with a few compounds having hundreds (or even thousands in the most recent years) of activity measurements. (B) Histogram of protein targets in BindingDB having a given number of activity entries. Approximately 400 targets were tested against one compound. Inset shows the fine-grained details of the distribution with approximately 500 compounds having been tested against two targets.
Reprinted with permission from Nicola et al.[25] Copyright 2012 American Chemical Society.

databases has been discussed and the importance of the complementarity between public and private databases has also been highlighted[5]. The role that "federated databases" such as Chemical And Biological Informatics NETwork (CABINET) can play in the future of the field has also been discussed by Olah and colleagues[14] as well as the key issue of data normalization and transparency among the different databases.

The trends and future directions of the combined holdings of BindingDB, ChEMBL, and PubChem have also been discussed with perceptive insights by Nicola et al.[25] They discuss the issues of coordination to minimize the duplication of effort and data, long-term sustainability related to funding sources and others. Future capabilities for the next generation of affinity databases are also discussed including the impact of the majority of the relevant literature becoming "interactive documents" allowing fingertip access to the available data for targets, pathways (via target databases such as supertarget database), and "omics" databases. Undoubtedly, a critical element of future developments would be the use of these vast amounts of data in a prospective and predictive manner to expedite drug discovery using increasingly more powerful computational tools.

An additional important consideration should be the quality of the existing affinity data (K_i, IC_{50}, or equivalent) in the different databases. This includes the clerical and transcriptional errors indicated by Gilson and colleagues[25] as well as the inherent limitations of the experimental data. The latter aspect has been analyzed recently by Kramer et al.[4] The authors estimate that the experimental uncertainty on the K_i measurements would result in a mean error of 0.44 pK_i units ($pK_i = -\log K_i$), a standard deviation of 0.54 pK_i units and a median error of 0.34 pK_i units[4]. Because pK_i is a log scale, these are relatively large errors (1 unit difference amounts to a factor of 10) that should be taken into account when attempting to extract general trends in the relationships between chemical structures and biological activities. This limitation would suggest that the choice of the most suitable variables to represent the content of these databases could play a significant role in our ability to extract meaningful relationships and inferences derived from relating the chemical and biological worlds.

Finally, one philosophical remark should be made as to the significance of the immense amounts of data contained in the SAR databases. Data are critical in the process of scientific discovery; this is true for any domain of scientific inquiry. From the historical perspective of the physical sciences, the astronomical data collected by Thyco Brahe (circa 1600) in the court of King Frederick II of Denmark were critical for the development of astronomy and the physical sciences. The Copernican notion of a Sun-centered universe needed to be put to a rigorous numerical test and the available astronomical data at the time were not accurate enough to test the hypothesis. It was only when superb data, using the best instruments of the time, were collected on clear nights on the Island of Hven (Oresund, Denmark) and were studied and patiently analyzed by J. Kepler (1571–1630) that a consistent and numerically reliable picture of the cosmos began to emerge. I have argued before[26] that our knowledge of the chemicobiological universe related to drug discovery is still rather limited when compared to the physical sciences. Our concerted efforts in the drug-discovery field, from the pioneer efforts of Paul Ehrlich (1854–1915) of Salvarsan fame in the twenty-first century, have continued for barely a century. The developments in computational and Internet technology have permitted the building of a number of extensive and interconnected databases relating CBS. However, the complexities of the biological world and the vastness of the chemical universe have prevented us from extracting the laws of "planetary motion" in the domain of drug discovery. Although large, the amount of data that we have relating the vastness of the chemical space to the immense and complex biological space is rather fragmented. We have identified some promising chemical classes of compounds and a relatively small number of biological targets (see Table 1.2, Figs. 1.2, 1.5A and B) amenable to regulation by our chemical ingenuity but we are still far from major encompassing generalizations. Vast amounts of data *per se* do not reveal whether the trajectories in CBS are circles or ellipses. One of the tenets of this work is that being able to provide a graphical representation of how the chemical and biological domains relate to each other would be a major step in the right direction. Unexpected insights could come from finding new variables to connect the two domains of molecular entities (chemical and biological) and from finding new ways to represent the myriad of data stored in the SAR databases.

CHAPTER 2

The Variables: Definitions of Ligand Efficiency Indices

The question of the suitability of the variables to study and analyze any experimental phenomenon has always been part of the scientific endeavor. What is(are) the most appropriate variable(s) to quantitatively describe and analyze a series of observations? The correct answer is always critical. Often, finding the correct variables takes us a long way toward understanding the critical parameters of physical phenomena. For example, Kepler formulated the description of the motion of the planets around the Sun rigorously in polar coordinates (r, Θ), a natural set of coordinates to describe elliptical motion. In turn, this description paved the way toward the concepts of central force and the corresponding Newtonian descriptions.

The dominant parameter in drug discovery has been the affinity between the target and the ligand. The introduction of the physicochemical properties of the ligand (e.g., Lipinski's Rule of Five, Ro5) came later in an "ad hoc" manner, imposing limitations on the choice of molecules. More recently, the appearance of "combined" variables such as Ligand Efficiencies (LEs) related to the size of the ligand have entered into the field but are still considered of limited value. I wish to suggest that finding the correct set of variables to guide the future of drug discovery is a critical step. In this chapter, we review the different formulations of combined variables that have appeared in the field and provide an analytical assessment of their relative merits.

2.1 LIGAND EFFICIENCIES BASED ON FREE ENERGY OF BINDING

It is probably appropriate to begin this section with a tribute to the seminal work of Kuntz and coworkers[30] and the lesser-known paper of Andrews et al.[29] for their initial insight into the affinity of ligands toward their targets. Kuntz's initial observation was that, for molecules containing up to approximately 15 non-hydrogen atoms (NHAs), the free energy of binding was proportional to the number of NHAs of the ligand. This insight contained the notion of a ratio, or efficiency of binding per atom, that they estimated to be approximately -1.5 kcal/atom[30].

For the purpose of this book, the initial numerical definition of ligand efficiency (LE) based on the free energy of binding was put forward in the brief note of Hopkins et al.[18] Their suggestion was simple but insightful. By dividing the ΔG of binding by the heavy atom count of the ligand, it should be possible to compare on a per atom basis different lead molecules (LE $= \Delta G$/HAC. See Table 2.1 for this and ensuing definitions of the initial variables). They also hinted at their use and the possible strategy of maintaining the value of the "LE" constant during subsequent efforts to optimize future lead compounds[18]. Most recently, the negative sign of the definition originating from the values of ΔG has been dropped and the LE derived from this definition is quoted as a positive quantity. This concept was relatively quickly accepted among medicinal chemists and is still the most commonly used[31-33].

The connection between affinity of chemical ligands toward their targets and its accurate calculation based on the fundamental of thermodynamics has always been in the mind of physical chemists. The initial paper by Hopkins et al.[18] as well as the drive to understand the relative contribution of enthalpy (ΔH) and entropy ($-T\Delta S$) to the binding energies has resulted in contributions from several researchers trying to identify the most reliable predictors of binding energies (enthalpic vs. entropic) during the optimization process. Freire[34,35] had emphasized earlier the accurate measurements of the two separate terms in various target–ligands pairs and suggested their use to rank the relative merit of the compounds in the optimization process. From the theoretical standpoint, other criteria and definitions of efficiency related to the two

Table 2.1 Initial LE Definition and LEIs Pairs Used in Each Efficiency Plane with the Corresponding Algebraic Description

Variable Name	Definition	Example Value[a]	Equation Number
LE	ΔG/NHAC	0.50[a]	(2.1)
BEI	pK_i, pK_d, or pIC_{50}/MW(kiloDaltons)	27	(2.2)
SEI	pK_i, pK_d, or pIC_{50}/(PSA/100 Å^2)	18	(2.3)

Slope of lines: 10(PSA/MW). Algebraic description*: BEI = 10(PSA/MW) SEI. Description: Efficiency plane based on macroscopic physicochemical properties of the ligand: PSA, MW

NSEI	$-\log_{10} K_i$/(NPOL)=pK_i/NPOL(N,O)	1.5	(2.4)
NBEI	$-\log_{10} K_i$/(NHA)=pK_i/(NHA)	0.36	(2.5)

Slope of lines: NPOL/NHA. Algebraic description: NBEI = (NPOL/NHA) NSEI. Description: Efficiency plane based on atomic properties of the ligand. Slope of the lines is always a rational number given as NPOL/NHA.

nBEI	$-\log_{10}[(K_i/\text{NHA})]$	10.25	(2.6)
mBEI	$-\log_{10}[(K_i/\text{MW})]$	11.5	(2.7)
NSEI	$-\log_{10} K_i$/(NPOL)=pK_i/NPOL(N + O)	1.5	(2.4)

Slope of lines: NPOL. Algebraic description:
nBEI = NPOL · NSEI + $\log_{10}$(NHA);   (2.8)
mBEI = NPOL · NSEI + $\log_{10}$(MW);   (2.9)
Intercept: $\log_{10}$(NHA) or $\log_{10}$(MW) respectively.

mBEI	$-\log_{10}[(K_i/\text{MW}]$	11.5	(2.7)
SEI	pK_i, pK_d, or pIC_{50}/(PSA/100 Å^2)	18	(2.3)

Slope of lines: 0.01 PSA. Algebraic description*: mBEI = (PSA/100) SEI + $\log_{10}$(MW). Intercept: $\log_{10}$(MW).

LLE (Ligand Lipophilicity index) = pK_i–log P (C log P or ACD log P)		(2.10)

log P: $\log_{10}$ of the octanol/water partition coefficient. C log P, ACD log P (log P values calculated by two different methods widely used in medicinal chemistry). PSA: Polar surface area or topological polar surface area. By substitution of any of the LE definitions above for pK_i (i.e., Eqs. (2.2)–(2.5) as the simplest examples) in Eq. (2.10), it is possible to have combinations of LLE and the corresponding LEIs above. For example, using Eq. (2.4) one can obtain:
LLE = NPOL · NSEI–Log P. This simple manipulation incorporates LLE into the variable framework. NPOL is defined as NPOL= Count of N + O.

Slope of lines: NPOL. Intercept: Log P. Combining LLE and NSEI in an acceptable atlas-related plane.

*Algebraic description given as the y-variable of a linear function of x (y = ax + b). a = slope, b = intercept. Mnemonic note: LEIs definitions beginning with a capital letter N (NSEI, NBEI) refer to a straight quotient between pK_i (=$-log K_i$, or corresponding affinity variable; Eqs. (2.4),(2.5)) and an integer number derived from the number of atoms of the ligand. Number of polar atoms (NPOL) for NSEI and number of heavy (nonhydrogen, NHA) for NBEI. The definitions of LEIs beginning with a lowercase letter (nBEI, mBEI) refer to variables in which the logarithm (base 10, log_{10}) operation is taken after the ratio of the corresponding variables (Eqs. (2.6), (2.7)). Because of this subtle difference in the definitions, the slope of the lines in the nBEI vs. NSEI (or mBEI vs. NSEI) plots is given by NPOL. Algebraic details have been published elsewhere[48] (supplementary information) and are easy to derive from the definitions given above (Appendix A).
[a]Adapted from earlier publications where example values for MW, PSA, NHA, and NPOL for a compound with a K_i = 1 nM were given[41,42]. The original definition of LE (Eq. 2.1) uses NHAC (nonhydrogen atoms); here NHAC = NHA and negative sign has been deleted. Approximate equivalences (scale factors) between the various size-related efficiency indices can be obtained by a direct ratio (i.e., BEI/LE ∼ 54). Similar values are obtained by algebraic manipulations in Appendix B.

separate components of the dissociation constant ($pK_D = pK_H + pK_S$) were also studied trying to observe trends during the optimization process. Related to Hopkins' LE but using only the enthalpy-dependent affinity, a size-independent enthalpic efficiency (SIHE) was defined by Ferenczy and Keseru[36] to compare compounds based on their thermodynamics of binding. The relative merits of enthalpic vs. entropic contributions, and their relative efficiencies, during the optimization process are still the focus of intense investigation using both experimental data on available databases[37] and from the theoretical standpoint[38] (and

Table 2.2 Extended Definitions of LEIs and Optimization Metrics[a]		
Variable	Definition	Equation Number
Size-corrected (or size-independent) LE definitions		
FQ	$$FQ = \frac{LE}{(0.0715 + (7.5328/(N_{Heavy})) + (25.7079/(N_{Heavy})^2) - (361.4722/(N_{Heavy})^3)}$$	(2.1)
%LE	$$\%LE = \frac{LE}{\left(1.614^{\log_2\left(\frac{10}{N_{heavy}}\right)}\right)} \times 100$$	(2.2)
SILE	$$SILE = \frac{-RT \ln(pK_i)}{(N_{Heavy})^{0.3}}$$	(2.3)
Efficiency indices related to lipophilicity		
LLE	$LLE = pK_i - \log P (\text{or } \log D)$	(2.4)
LLEAstex	$$LLE_{Astex} = \frac{0.11 \times \ln(10) \times RT(\log P - \log(K_d \text{ or } pK_i \text{ or } IC_{50}))}{N_{Heavy}}$$	(2.5)
LELP	$$LELP = \frac{\log P}{LE}$$	(2.6)
Efficiencies related to enthalpy		
EE	$$EE = \frac{\Delta H}{N_{Heavy}}$$	(2.7)
SIHE	$$SIHE = \left(\frac{-\Delta H}{40 \times 2.303 \times RT}\right) \times (N_{Heavy})^{0.3}$$	(2.8)
More complex metrics only indirectly related to LEs		
MPO	Log P, log D(pH 7.4), molecular mass, tPSA, H_{don} and pK_a	
CSE	*In vitro* promiscuity and toxicity data, log P, tPSA and pK_a	
DRUG$_{eff}$	$$DRUG_{eff} = \frac{\text{biophase concentration}}{\text{dose}} \times 100$$	(2.9)
QED	$$QED = \exp\left(\frac{\sum_{i=1}^{n} w_i \ln d_i}{\sum_{i=1}^{n} w_i}\right)$$	(2.10)

[a]*LE: as defined in Table 2.1 Eq. (2.1)[18], FQ: fit quality index[53], %LE: ratio of LE to maximum LE as defined by Orita et al.[53], SILE: size-independent ligand efficiency[52], LLE: lipophilicity ligand efficiency[54], LLE: lipophilic efficiency (Astex definition)[55], LELP: ligand efficiency log P[56], EE: enthalpic efficiency[57], SIHE: size-independent enthalpic efficiency[36], MPO: multiparameter optimization related to central nervous system (CNS) compounds[58], DRUG$_{eff}$: estimation of the in vivo efficacy from the in vitro potency[59], QED: quantitative estimate of drug-likeness[60]. Adapted from Hann and Keseru[61].*

references therein). The multiple definitions of the various extensions of the original variables related to LE are listed in Table 2.2. The interest in understanding the thermodynamics of protein–ligand binding from the basic principles and how these elements contribute to drug discovery will continue. In particular, the separation of enthalpic and entropic terms and even more importantly the hydrophobic vs. polar (and solvent derived) effects is critical. However, in what follows we would emphasize the simplest definitions of LEs, based on the affinity constants (K_i, K_d, IC_{50}) and their relationships to size (e.g., MW, NHA) and polarity (number of polar atoms or polar surface area (PSA)).

From a practical standpoint, other interesting variations on the concept of efficiency were developed later. In particular, the concept of "group efficiency" (GE) as a guide to optimize the hit-to-lead process has also been put forward by the Astex group[39]. It refers to the change in free energy of binding ($\Delta\Delta G$), added to an existing molecule "A" to form molecule "B," scaled by the additional NHAs provided by the adduct (ΔHAC). This concept has been found useful by certain laboratories[31,39] and may be useful for the optimization process when certain chemical groups or scaffolds remain constant. It can also be used to assess the contribution of a certain "chemical group" to the efficiency of the whole. More data are needed to assess the value of these concepts in the wider drug-discovery community.

2.2 LIGAND EFFICIENCIES BASED ON PHYSICOCHEMICAL PROPERTIES

An expanded definition of the original concept proposed by Hopkins et al.[18] was suggested soon thereafter that included not only the efficiency per size but also the efficiency per unit of PSA, as a measure of polarity and hydrophilicity (or polarity). It is possible that the inclusion of the broad concept of PSA into an efficiency metric might seem a bit implausible to some, but the idea was to compare molecules based on their polarity. In this context, PSA seemed a good starting point, as a complementary variable to MW. Abad-Zapatero and Metz[40] introduced a pair of ligand efficiency indices (LEIs) (BEI, SEI) as two complementary variables in approximately the same scales, which allowed a comparison of the chemical entities in two dimensions and also permitted a graphic representation in 2-D plots, typically (SEI, BEI) as (x,y). Algebraic definitions for all the LEIs described in this section and throughout this work are listed in Tables 2.1 and 2.2 for convenience.

An important point that was not fully discussed or appreciated in the paper by Abad-Zapatero and Metz[40] was that in this representation, the ratio of BEI/SEI for any target–ligand pair, which is the slope of the lines in the plots, is equal to 10(PSA/MW) and therefore depends only on the physicochemical properties of the ligand. This is obvious given that the definitions of these two variables share the numerator $(-\log K_i = pK_i)$ and the two scale factors for the units of MW and PSA differ by a factor of 10. These insights were fully discussed in a later paper[41] that also illustrated the use of these two variables to represent the various compounds on the (SEI, BEI) planes for two targets: human protein tyrosine phosphatase 1B (hPTP1B) and human fructose bisphosphatase (hFBPase). In addition, the limited data examined at the time suggested that all the chemical entities for various targets could be enclosed (or bounded) by a wedge with the corresponding range of PSA/MW values. This prediction was later confirmed using the SEI–BEI efficiency plane to map the content of PDBBind (Fig. 2.2A). Further details can be found in reference[42] (Supp. Material).

Following these publications, some groups[43] began using the value of BEI as a valuable index to monitor their optimization efforts, most typically to move from initial hits to lead compounds for further development[44,45]. Significantly, the values of the complementary variable SEI were typically not followed in the optimization process. It appeared that the most important (and almost only) factor in the optimization process was potency in relation to size[46]. Only the ligand lipophilicity index (LLE) (see Table 2.2) attempted to include hydrophobic effects in the optimization criteria and it is also commonly used[33].

Importantly, the work of Tanaka and colleagues[47] presented a detailed prospective application of the use of a LE-guided process to design soluble epoxide hydrolase inhibitors. This work explicitly used the values of several LEIs to monitor their progress (LE, BEI, SEI) and in the process they suggested a sound strategy. A few other insights and findings of this work are worth summarizing here.

The authors present a distinct correlation between BEI and the initial LE defined by Hopkins et al.[18] (Fig. 2.1). Although it was to be expected, Tanaka et al.[47] presented the correlation between the values of BEI and LE for their initial hits and how they used these values to select the most promising compounds for development based on their optimal efficiency indices (LE > 0.37, BEI > 19.5). Their initial compound (compound 1) was

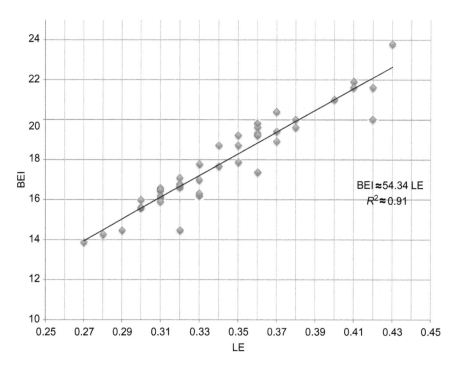

Figure 2.1 Relation between LE and BEI. The scale conversion between BEI and LE obtained by least squares regression (N = 42). Data kindly provided by D. Tanaka[47]. See Appendix B for an algebraic derivation of the approximate value of this scale factor.

a hit of high efficiency (LE = 0.43, BEI = 23.8, SEI = 11.8). This compound was not the most potent and was purposely selected using the most favorable LEs values. The project was able to successfully achieve a potent, ADMET-acceptable lead-like compound (**11**) by rapid medicinal chemistry of high efficiencies (LE = 0.43, BEI = 22.8, SEI = 11.3). The work did not use any screening in the conventional sense but rather was driven by a LE-guided hit triage of a typical virtual screening effort monitoring a wide range of "LE" criteria including the conventional definitions and the size-corrected definitions such as FQ and SILE discussed below and in Table 2.2.

2.3 LIGAND EFFICIENCIES BASED ON ATOMIC PROPERTIES

There was a decisive interest in developing other LEIs that would allow comparison of the chemical entities on the basis of their atomic composition. An obvious extension, along the lines suggested by Hopkins et al.[18], was to divide the affinity variable (typically expressed as pK_i) by the heavy atom count (HAC = NHA, for this discussion) to introduce an atomic, size-related, binding efficiency, called NBEI. Similarly, dividing the affinity variable by the number of polar atoms (NPOL) did provide a measure of binding efficiency per polar atom count: NSEI. It should be clarified that the value of NPOL selected in the original definition was simply based on counting the number of N and O in the molecule, without taking into account the specific chemical environment of these atoms. This definition is clearly simplistic from the chemistry viewpoint but helps in providing the "general chemical features" of the ligand. This atomic-based definition of LEIs also permits a graphical representation of the ligands and their corresponding targets (connected by their affinity variables: K_i, IC_{50}, K_d) in efficiency planes (NSEI, NBEI). In this representation, the slope of the lines is given by the ratio of NPOL/NHA and therefore the polarity of the compounds increases counterclockwise in the diagrams. As before, the position of the individual target–ligand pairs is given by the radial coordinate along the line with slope NPOL/NHEA. Consult Table 2.1 for a concise algebraic definition of the terms and the appearance of the corresponding plots.

In the process of analyzing the results of this new definition of LEIs based on the atomic composition of the ligands, it was discovered that a small variation in the algebraic definition of the LEI referring to size (NBEI), could provide a very appealing and easy to interpret series of plots. This new definition of efficiency per NHA count (nBEI) was based on taking the negative $\log_{10}$ of (K_i/NHA) (see Table 2.1 for algebraic details). It can be shown[48] (Appendix A) that plots of nBEI vs. NSEI are described by a family of lines given by the expression in Eq. 2.11 (Fig. 2.2A–B):

$$nBEI = NPOL \cdot NSEI + \log_{10}(NHA) \qquad (2.11)$$

In other words, the slope of these lines is given by the number of polar atoms, (NPOL = N + O, as defined above) and the intersect of those lines is the $\log_{10}$(NHA). Once again, after algebraically eliminating the dependence of the affinity constant, the linear position of the compounds on a 2-D plane can be predicted only on the basis of the atomic properties of the ligand (slope and intercept). The appearance of these plots is illustrated in Fig. 2.2B, for the approximately 1,300 compounds included in PDBBind (2007 release, Fig. 1.1).

(A)

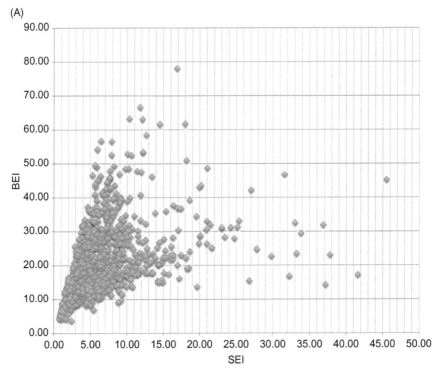

Figure 2.2 (A) Overview of the content of PDBBind (V2007) represented in the SEI–BEI efficiency plane. This representation was one of the first results of applying combined variables to map the full content of a database relating targets and their ligands. The approximately 1,300 target–ligand pairs were extracted from the PDB 2007 release and curated as indicated in Fig. 1.1, including the calculation of SEI and BEI. Although the distribution of available ligands in PDBBind is restricted by the limitations of the protein crystallography methodology, the database includes an approximately random distribution of affinities and polarity (Fig. 1.1). The first is reflected in the plot by the uniform distribution of points along the radial coordinate of each line, and the second by the distribution of points in the angular coordinate. The isolated points mapping at SEI > ~25 correspond to the few very potent and very hydrophobic ligands contained in PDBBind. The remaining wedge of target–ligand complexes (one per diamond) represents the full content of the database.

Similar considerations using the variable mBEI, defined in Table 2.1 (Eq. 2.7), give a family of lines:

$$mBEI = NPOL \cdot NSEI + \log_{10}(MW) \qquad (2.12)$$

Thus, any chemical entity can be placed on a line defined by the above NPOL and intersects. If the compound(s) have any activity against any biological target (i.e., an enzyme or receptor) then the target–ligand pair can be mapped unambiguously on the plane. Values of affinity against different targets will slide the position of the target–ligand pair along the line. Further details of these concepts and representations have been presented recently[48,49].

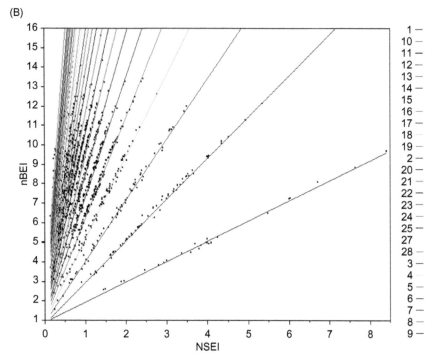

Figure 2.2 (B) Overview of the content of PDBBind (V2007) represented on the NSEI–nBEI efficiency plane. This plane represents the same data as Fig. 2.2A but using a different pairwise set of variables NSEI and nBEI. Although conceptually both Cartesian planes use similar variables (polar-related x, size-related y), the subtle differences between them (Table 2.1) provide a more discrete and appealing distribution of target–ligand pairs. The lines correspond to the regression lines modeled statistically with slopes equal to NPOL (N + O count) as color-coded in the right panel and increasing counterclockwise from 1 to 28. The ligands along each line contain the corresponding number of polar atoms (N + O) given by NPOL. The distribution of points along each line corresponds to the approximately random distribution of K_i included in PDBBind. The same thing can be said for the number of points for each angular coordinate. Details of the statistical analysis are given in the supplementary material of Reference[42].

The definition of NPOL as the N + O count was done initially for simplicity and indeed has limitations. However, it relates well to one of Lipinski's Ro5[50] (see below and Part II). A much more fine-grained description of the polarity of molecules (e.g., using PSA calculations based on the 3-D conformation), combined with other pairs of variables (BEI, SEI) in efficiency space can help solve this problem. Other alternatives could be the distinction between different types of N and O in the SMILES (2-D or 3-D) representation of the chemical compound, given their chemical environment, in order to fine-tune the representation of the molecules in the various efficiency planes (see, for example, mBEI, SEI at the end of Table 2.1). These refinements could be incorporated in subsequent elaborations of the "polarity-related" efficiency indices.

Returning briefly to the LLE mentioned above, it is easy to see how the LLE index can also be readily incorporated into 2-D planes as follows. From the definitions of LLE and the atomic polar surface efficiency index NSEI given in Table 2.1:

$$NSEI = pK_i / NPOL \qquad (2.13)$$

It is possible to substitute the pK_i on the definition of LLE to yield:

$$LLE = NPOL \cdot NSEI - C \log P \qquad (2.14)$$

This implies that plots of LLE vs. NSEI will also show lines with slopes NPOL, making them easy to interpret (see Table 2.1).

2.4 A FORMULATION IN POLAR COORDINATES

In view of the suggested interpretation of the SEI, BEI (x,y) planes in terms of polar coordinates (see Table 2.1, Fig. 2.2A), a more rigorous presentation is given. It is preferable to select SEI as the x-coordinate so that the value of PSA/MW increases counterclockwise (positive slope) in the plane.

If we define the modulus of the LEI for BEI and SEI as:

$$LEI = \sqrt{(SEI^2 + BEI^2)} \qquad (2.15)$$

then the two components of the vector refer to "binding-per-PSA" (x, SEI) and the "binding-per-MW" component (y, BEI). The corresponding angle in polar coordinates will be given as: $\theta = \arctan [10(PSA/MW)]$. With this definition, the two components of the modulus of the LEI vector are given by:

$$x = SEI = LEI \cdot \cos(\theta) \qquad (2.16)$$

$$y = BEI = LEI \cdot \sin(\theta) \qquad (2.17)$$

with θ defined as above, increasing counterclockwise.

This formulation can be applied to any dataset containing an affinity value (preferably K_i) related to a consistent experimental enzymatic assay under similar conditions for all the compounds tested. This interpretation might be considered a far-fetched extrapolation, but in the future it could offer a way to compute Cartesian distances between target−ligand complexes to provide an optimization parameter, when compared with existing milestone (i.e., drugs) compounds. In addition, this distance in efficiency space could also be used to more rigorously define the specificity of a compound toward two different targets.

Thus, if the angle theta is approximately 45° for any compound or series of compounds, then the corresponding chemical entities will map approximately along the diagonal. This will occur for compounds for which the PSA/MW is approximately 0.1, as 10(PSA/MW) will be approximately equal to 1.0, and the two variables might "appear" to show a correlation along the diagonal of the plane. Independent of the value of their angular coordinate, compounds belonging to different chemical series or projects will appear to be highly correlated (lying approximately along straight lines) if their PSA/MW ratios are very similar. However, this can be achieved by different chemical compositions and structures. It is not the purpose of this representation to make statistical inferences but to map in two dimensions how the chemical entities relate to their target via the affinity parameters (K_i, IC_{50}, K_d).

The strength of this representation resides in its simplicity. Future extensions of the mapping of chemicobiological space (CBS) could include the expansion of any line or wedge (in the SEI−BEI plane) into a separate part of chemical space where the structures and chemical characteristics of the specific compounds can be analyzed further. Similarly, in the NSEI−nBEI plane, compounds (lines) with a given value of NPOL can be represented into other efficiency planes by the appropriate change of coordinate axes and scales. Naturally, there could be many compounds in between the lines corresponding to NPOL=1, 2, etc. depending on the $\log_{10}(NHA)$ intersect (Table 2.1). The changes in the $\log_{10}(NHA)$ term will make the appearance of the lines "thicker" when there are many compounds with the same NPOL number and a large number of chemical substitutions or variants within that same scaffold (see, for instance, Fig. 2.2A and B, Fig. 3.2 and examples in Parts II and III).

2.5 REFINEMENTS TO THE DEFINITIONS OF LIGAND EFFICIENCY (LE)

Not surprisingly, the initial definition of LE introduced by Hopkins was soon revised to accommodate larger datasets, including bulkier compounds. Since the original work of Kuntz and coworkers[30], it had been noticed that the binding affinity of ligands for their targets was dependent on the size of the ligand and reached a plateau at approximately 20–25 NHAs. The definition of LE introduced by Hopkins in 2004 provided a simple algebraic framework to address this issue in a straightforward manner. Datasets larger than that originally used by Kuntz permitted numerical corrections to the simple ratio of affinity to atoms. Thus, size-independent (or size-corrected [SILE]) LE values have been presented in the literature basically in three different formulations (Table 2.2). In relation to the original LE parameter, the simplest correction is given by:

$$SILE = -RT \ln(pK_i)/(NHA)^{0.3} \tag{2.18}$$

Additional analysis of the concept of LE has explored further what was initially hinted at by the seminal work of Kuntz and collaborators, namely that the efficiency of a ligand is size dependent[30]. The observation that on average smaller ligands have greater efficiencies than larger ligands has now been amply documented[51], with a dramatic decrease between 10 and 25 NHAs. They proposed two causes for this effect. Firstly, deterioration of the quality of fit between the ligand and the pocket within the target, as the ligands increase in size. Secondly, reduction in the PSA of the ligand (on a per atom basis), as the size increases. Although the first cause seems most likely correct, the second one can be questioned for large flexible ligands such as peptides.

To correct for this effect, Reynolds and colleagues have introduced the "fit quality score" (FQ score) that scales the LE by a cubic polynomial, which fits the observed size dependency (using HA count) as a variable[51]. A simpler scaling has been suggested recently by Willem and Nissink[52] of the form: SILE = affinity (pK_i, pIC_{50})/NA. Significantly, the authors show the effect of using uncorrected LE vs. SILE in monitoring the progress of fragment-design efforts (Fig. 5 in reference[52]).

The other two size-independent corrections also have in the denominator the variable related to the number of NHA (Table 2.2). This would suggest that an algebraic equivalence could be found between the three. An approximate scale factor can be found by simple regression analysis among the three separate values. The correlation coefficients and equivalences are: FQ ~ 0.0134%LE ($R^2 = 0.951$, $N = 42$), FQ = 0.365 SILE ($R^2 = 0.962$, $N = 42$) and correspondingly %LE = 27.1 SILE ($R^2 = 0.999$, $N = 42$) (Fig. 2.3A–C).

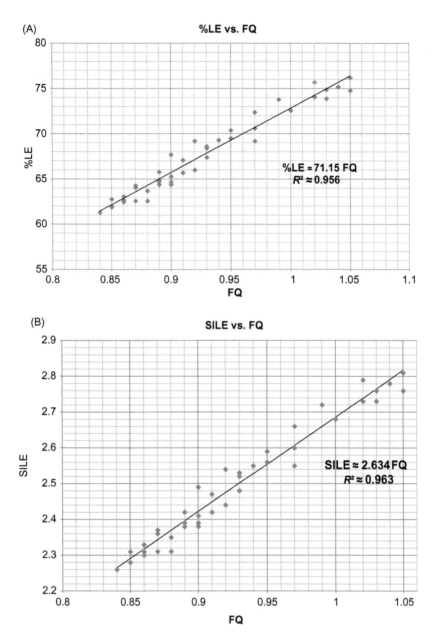

Figure 2.3 Equivalences between different size-independent LEs. (A) Linear regression between %LE and FQ. Least squares fit between two of the size-independent LEI listed in Table 2.2. Although in two very different numerical ranges the two different size corrections are correlated. For certain targets and under various assay conditions there might be local differences, but the two variables are basically equivalent. A scale factor could possibly also be found by numerical and algebraic manipulations. (B) Linear regression between SILE and FQ. As above, least squares analysis between two (SILE and FQ) of the size-independent LEI listed in Table 2.2. The result is similar to the one obtained previously except for the values on the y-axis and the corresponding scale factor. The relationship could possibly also be found by numerical and algebraic manipulations using the definitions.

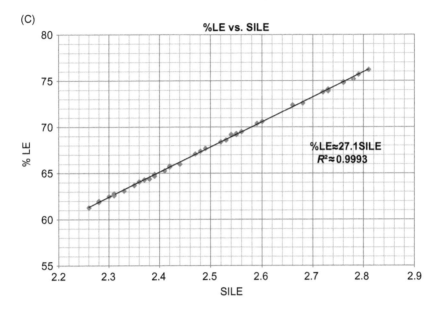

Figure 2.3 (C) Linear regression between SILE and %LE. This is the best least squares regression resulting from the fit between SILE and %LE, strongly suggesting that they are completely equivalent, aside from minor local variations in the affinity measurements. Data kindly provided by D. Tanaka[47].

There is little doubt that the size-corrected LE values better describe the available data than the straight ratio originally proposed in the corresponding publications. However, those improvements amount to small corrections when considering the uncertainty in the values of the affinity values themselves (see Chapter 1). Thus, although we recognize the future importance of the best possible LE efficiency parameters, the current formulation of a unified framework utilizes only the simplest, "first approximation" LE values. Refinements on the efficiency variables will be incorporated when needed.

2.6 DISCUSSION

From the initial work of Kuntz and collaborators[30], the concept of referring the activity of ligands to their physicochemical properties has expanded in several directions providing different quantitative ways of scaling the affinity of the ligands by their corresponding physicochemical properties. Currently, there are LEI related to ΔG, to the affinity constant themselves (K_i, IC_{50}, K_d), to lipophilicity, enthalpy, and other variables. In addition, LEIs can be defined in relation to size- and polarity-related properties of the chemical entities. Perhaps because of the difficulties in defining the polarity-related quantities, LEIs related to polarity have been less common in the recent applications of these ideas to practical drug-discovery projects.

The summary presented here shows that most of the size-related LEIs are equivalent (except for a factor of scale, see Appendix B) and can be used independently of each other (or possibly in combination in certain cases) as guiding variables to select compounds for further optimization. Size-independent corrections are basically equivalent and represent a small correction to the predictive value of the calculated efficiencies. This correction could play a role in the future but currently the uncertainties in the experimental (and theoretical) values of the affinities tend to diminish the importance of these numerical corrections.

Only a few polarity-related LEIs have been defined (e.g., SEI, NSEI), including those related to lipophilicity (LLE, LLE$_{Astex}$, LELP) and possibly the enthalpic efficiencies (EE, SIHE) when the binding energies can be related to this physicochemical potential (ΔH). Polarity/lipophilicity are more difficult parameters to assess and yet they are critical for the issues of solubility and pharmacokinetic properties. However, it is also essential to find the correct combination of variables to incorporate these factors into a robust formulation for drug discovery. An initial mapping of the content of a database (PDBBind) in terms of two complementary LEIs (polarity, size: x,y) has been given, providing a way to represent the content of CBS (Fig. 2.2A and B). We suggest that, in the long term, it will be the insightful combination of size- and polarity-related LEIs, as a first approximation, that will have the stronger predictive value. Naturally, additional variables and refinements to this approximation will increase the predictive power of these numerical or statistical formulations.

Variables and Data: The AtlasCBS Concept

The latitude and longitude of geographic locations *per se* are not very visual. Similarly, the list of *x,y* coordinates of the points that make a circumference is not very revealing. It is only when the first are represented to make a map and the latter ones are plotted on a plane to establish the contour of a circumference that we have the insight, the "aha" moment, that extracts the concept from the list of numbers. For further detail, we can go to the individual list of coordinates or points in the plane to quantitatively assess distances and references. However, the overall grasp of totality comes from an adequate graphical representation. We suggest that a certain formulation of the efficiency indices related to size and polarity, used in a pairwise combination, can provide a graphical representation in planes (i.e., efficiency planes) that could be considered akin to an atlas-like representation of chemicobiological space (CBS).

3.1 THE VASTNESS OF CHEMICAL SPACE

An inspiring review by Lipinski and Hopkins[62] compared the vastness of chemical space to the cosmological universe, with the chemical compounds populating the space in the same way that stars fill the cosmos. The number of possible chemical entities derived from even the simplest chemical scaffolds is unimaginable: $\sim 10^{29}$ derivatives of *n*-hexane. Within this vastness, the challenge for chemical biologists and drug explorers is to find the subset of biologically active molecules that can affect biomedical targets therapeutically and as modulators of biological systems.

In their vision, the authors portrayed the chemical space as a blue continuum with embedded ellipsoidal volumes of various sizes occupying certain regions of space where molecules of specific affinities would map. Thus, "islands" of compounds active toward aminergic G protein-coupled receptors (GPCRs), proteases, kinases, and other macromolecular targets would exist within this space as "galaxies" overlapping to a certain extent with a central volume that would contain the subset of molecules exhibiting drug-like properties. These were defined as chemical entities possessing pharmacokinetic properties suitable for oral administration[62].

For obvious reasons, the schematic representation of the vastness of chemical space as it relates to biological space attempted by Lipinski and Hopkins was presented in three dimensions. However, it could be that the best way to relate chemistry and biology for the purposes of effectively finding active molecules is via a multidimensional space ($n > 3$) using variables still unknown to us, combining affinity to the targets with the physicochemical properties of the ligands. An extension and more concrete depiction of CBS is suggested in this section and illustrated in Fig. 3.1.

Given this immensity, concepts and tools have been developed to navigate this diversity. The concept of a "chemical positioning system" (i.e., "chemography") was proposed based on mapping the compounds using chemical descriptors as coordinates of different physicochemical or topological properties[63] that do not have to be restricted to three dimensions. However, this concept did not consider the use of the affinity constant relating the chemical entities to their corresponding targets and addressed only chemical space.

The subset of biologically active molecules toward their corresponding targets is "labeled" by a measureable affinity constant between the two entities. The measurements can be made *in vitro*, under well-defined conditions, or as an *in vivo* effect by the corresponding constants. The most reliable affinity constants between well-characterized chemical entities and assay conditions are typically K_i, IC_{50}, and K_d, and these are the critical links between the unique chemical matter and the target of choice in the corresponding assay. The physicochemical properties of

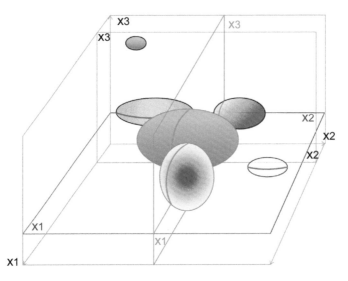

Figure. 3.1 *The vastness of chemical space and its relation to biological targets. A more concrete and expanded depiction of the vision proposed by Lipinski and Hopkins[62]. For convenience, CBS is represented in a 3D Cartesian system with axes x1, x2, and x3. This 3D representation is of a more complex n-dimensional space. Within this space, bounded regions (ellipsoids of different colors and sizes) might enclose the chemical entities active against certain targets (i.e., proteases, kinases, and GPCRs) and conceivably an enclosed region (central blue ellipsoid) could contain the subset of chemical substances with favorable "drug-like" properties. The different ellipsoids of active molecules against separate targets may (or may not) intersect with the drug favorable region, making those targets or "pockets" within the targets druggable or not. Sections, or planes, within this n-dimensional space (e.g., x1, x2 red; x2, x3 blue; and x1, x3 green as shown) could provide "map-like" representations of sections of CBS. The ensemble of maps focusing on different planes, regions of the map, various targets, different scales, and with different variables would constitute an atlas-like representation of CBS.*

the chemical matter, no matter how vast, are very well defined and include molecular weight (MW), nonhydrogen atom count (NHA), number of polar atoms (NPOLs), polar surface area (PSA), log *P*, and others, which have been shown to be critical for a therapeutic effect. Any attempt to map and effectively navigate the vastness of CBS has to take these biological and chemical variables into consideration when trying to devise the best "latitude" and "longitude" coordinates that will permit us to chart this complexity. Accurate mapping is required for safe navigation.

3.2 A UNIFIED FORMULATION

As discussed above, there have been various definitions of "size-related" efficiency indices using MW, number of NHAs, and other variables combined with the affinity measurement given in units of ΔG, or simply with K_i, IC_{50}, or the dissociation constant K_d. Fewer definitions of variables relating the potency of the ligand to its polarity are available (ligand lipophilicity efficiency [LLE] and surface efficiency index [SEI]), and undoubtedly each one has its own merits within the various drug discovery research laboratories, be they private or academic. Our motivation and interest in using a particular definition of the two most dominant variables (size and polarity) in terms of ligand efficiency index (LEI) was not partisan; it was focused on developing a unified view of the variables that drive drug discovery in similar and compatible scales that could (i) help in visualizing the drug discovery process, (ii) provide a more robust numerical framework for future optimization strategies, and (iii) possibly be extended with additional variables, providing an *n*-dimensional path for future numerical or statistical optimization. Technical and logical reasons for our choice of variables are given in the following paragraphs.

The definition of ligand efficiency (LE) by Hopkins et al.[18] has value in that it is related to the free binding energy (ΔG) between the ligand and the target. It has become very popular in the drug discovery and medicinal chemistry publications[33]. However, it presents several restrictions that should not make it the only choice. A few limitations of this definition are that it (i) treats the contribution of all atoms the same way (C vs. the heavier elements Cl, Br); (ii) cannot be used rigorously when affinity values other than K_i are available (IC_{50}, a very common measurement); and (iii) introduces the temperature factor, which in many assays is ambiguous or left undefined. Thus, an alternative definition of size-related efficiency was used for scaling the commonly accepted parameter for the affinity pK_i (or equivalent) to MW (binding efficiency index [BEI]). A similar complementary polarity-related efficiency was defined (SEI) by scaling the same affinity value to the PSA (see Table 2.1). These two complementary efficiency variables represent a unified framework for the concept developed later.

In addition, it is appropriate to emphasize that several of the definitions of size-related LEI are comparable in that they only differ by a numerical constant factor and thus their numerical predictive value is practically equivalent and their usage as mapping values results in a change of scale in the corresponding axis. A couple of examples will be detailed briefly and finding similar relationships for other pairs is left as an exercise.

In the work of Tanaka and coworkers mentioned earlier[47], they plotted the values of BEI and LE (Eqs. (2.1) and (2.2); see Table 2.1 for definitions) for the corresponding compounds of their discovery effort toward inhibitors of soluble epoxide hydrolase (Fig. 2A, $R^2 = 0.913$ of Tanaka and coworkers). Using their data (courtesy of D. Tanaka), it is possible to obtain a linear regression given by BEI ~ 54.34 LE (Fig. 2.1). A similar scale factor between the two can be obtained algebraically using the corresponding definitions of the two variables (Eqs. (2.1) and (2.2); Table 2.1), taking into account the appropriate units and the approximation suggested by Hopkins (MW ~ 1.32 NHA) for the MW of a typical, small medicinal chemistry molecule[18]. The simple algebraic manipulations are shown in Appendix B to yield BEI ~ 54.8 LE ($T = 300$ K). A more accurate equivalence can be obtained from the definitions of NBEI (Eq. (2.5)) and LE (Eq. (2.1)) (Table 2.1) as the denominators of both definitions are the same and no approximation is required. Appendix B provides the details showing that LE ~ 1.37 NBEI. Other equivalences (BEI, NBEI, etc.) are left as an exercise. Therefore, it is easy to deduce equivalence factors between the different size-related efficiency indices and conceptually they should be considered equivalent except for a change in scale. In certain cases, the predictive value of the different size-related indices to optimize a series could be different but, in our view, the differences should be cautiously considered in relation to the variability of the affinity measurements and assay conditions.

Except for the LLE (ligand lipophilicity index), no other formulation of ligand efficiency related to polarity has gained popularity in the community if one considers LLE "complementary" to SEI (or inverse of polarity). Equation 2.14 (Chapter 2) and note in Table 2.1 could provide a framework to include LLE into the unified formulation suggested below. Thus, we consider that the variables defined in Table 2.1, taken in pairs as indicated, provide a reasonable initial, analytical, and graphical framework to represent the contents of CBS in planes. Even if the fully effective description of CBS does require (and it most likely will!) additional variables, the concept of describing an n-dimensional space in 2D or 3D partial views (projections or sections) has value and will provide valuable insights to navigate more multidimensional spaces.

Further, refinements of this concept will most likely develop in the future. The schematic view is depicted in Fig. 3.1.

3.3 THE CONCEPT OF AN ATLAS-LIKE REPRESENTATION

Previous discussion and examples have shown how it is possible to represent the content of PDBBind in Cartesian planes using various definitions of LEIs. The image is simple and effective in representing the physicochemical properties of the ligands on the angular coordinate and the biological affinity along the radial coordinate. In addition, the choice of the Cartesian axes can reflect physicochemical properties (SEI, BEI), atomic properties (NSEI, NBEI) or other combinations (NSEI, nBEI) suggesting that each plane offers different insights or views.

It is probably useful to review the characteristics and appearances of each of these three "efficiency planes" in relation to the chemical and biological entities that they represent. Each point in the plane represents a target–ligand complex whose angular coordinate (or position along a concrete line) depends only on the properties of the ligand. In the efficiency plane SEI–BEI, the ratio 10(PSA/MW) of the ligand physicochemical properties determines the slope of the line, where the compound is and the exact position is given by the affinity value K_i (or equivalent) that defines the x,y components. In contrast, in the efficiency plane NSEI–NBEI, the slope of the line that the compound occupies is given by the ratio NPOL/NHA, which always yields a rational number.

The efficiency plane NSEI–nBEI (or the equivalent one NSEI–mBEI) is special in that the slope is defined strictly by NPOL(N + O) of the ligand and the intercept of the line is given by $\log_{10}$(NHA) (or $\log_{10}$ (MW)). Because the intercept depends on the logarithm of a relatively small number (NHA: 10–40, for typical compounds), the position of the point departs only very slightly for the different lines and the compounds cluster very conveniently along clear lines (Fig. 2.2B). Statistical analysis of the deviations reveals that for the PDBBind set, 98% of the variance is explained by NPOL (see Abad-Zapatero et al.[42] supplementary material). This brief summary highlights the striking effect that different variable definitions can have on the appearance and interpretation of the different efficiency planes.

The data currently available in the structure–activity relationship (SAR) databases discussed in Chapter 1 amount to our observations relating the chemical and biological domains of CBS. We surmise that the concepts related to LEI introduced earlier, the algebraic framework, and the resulting graphical representation can provide a sound basis to open novel insights into the ways in which chemical entities interact with their corresponding biological targets. The concept that we wish to put forward is the notion of a representation of CBS in an atlas-like form, using different pages (planes or maps) at different scales, using different variables and different portions of CBS specified by two variables or coordinates, that could metaphorically be related to latitude (BEI-like) and longitude (SEI-like), in conventional maps. The variables are based on various definitions of LEI related to two critical physicochemical properties for drug discovery, namely, size and polarity. The size variable can be basically given in two forms: MW and the number of non-hydrogen atoms (NHAs). The polarity variable is probably more controversial and we use PSA, topological tPSA, and number of polar atoms (N + O count).

We propose the use of these two complementary variables as Cartesian axes in a series of maps to graphically represent and relate the targets to their active small ligands. In the geographical analogy, these different maps, at different scales could be compared to the maps of a country, a state, or a city using physical features (rivers, mountains) or streets and directions (N, S, E, and W) as orientation aids. If an effective description of CBS for the purposes of drug discovery requires more variables, then appropriately selected planes (or sections) can convey the various dimensions and the different regions of CBS relevant to targets or chemical series or scaffolds. Whether or not these regions can be bounded or not (as suggested by Fig. 3.1) it is unknown at this time.

As indicated above, the suggested complementary variables (SEI-like and BEI-like) are appealing because: (i) the simplicity of their calculations (Table 2.1); (ii) they allow a separation between the ligand-related properties that are represented by their *angular* coordinate, and their relationship to the different targets, which is reflected in the *radial distance* from the origin. Compounds can be placed along a line using their chemical properties (NPOL or PSA/MW) and their unique position along the line is defined by their corresponding affinity value toward each target tested. We illustrate this initial concept by

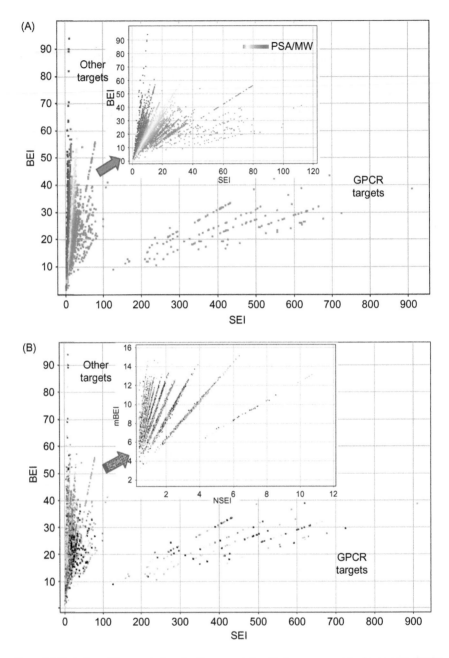

Figure. 3.2 Illustrations of the concept of an atlas-like representation. The images represent the drugs contained within WOMBAT (release 2007): 7,722 affinity measurements (3,585 K_i and 4,137 IC_{50}) using the ligand efficiency variables SEI and BEI. (A) Note the counter clockwise increase of polarity (PSA/MW from green to red) in the angular coordinate. Inset is a close up of the ligands active against non-GPCRs targets using the same variables. (B) Uses the same representation in the large efficiency plane but the inset corresponds to a change to the variables NSEI−mBEI, where clear lines of slope NPOL are recognized. These initial 2D views of the content of WOMBAT can be related to the depiction and mapping of CBS presented in the previous Fig. 3.1. Colors correspond to different targets. More details can be found in the original publication. Reprinted with permission from Abad-Zapatero et al.[42]. Copyright Drug Discovery Today.

representing the content of WOrld of Molecular BioActivity (WOMBAT) (v.2007) in Fig. 3.2, which demonstrates how it is possible to separate the regions of CBS containing ligands for GPCR targets from the ones relating to other targets. Further details of this concept were presented in a recent publication together with several conceptual applications to various aspects of drug discovery that will be discussed later in Part II. Reviewing these examples will help to better grasp the concepts presented.

3.4 AtlasCBS vs. OTHER NAVIGATION TOOLS IN MEDICINAL CHEMISTRY

Not surprisingly, the metaphors of geography and navigation when describing the relationship between chemical and biological space are not new. The concept of "chemography" (i.e., ChemGPS, the art of navigating in chemical space) was proposed more than a decade ago[63]. The analogy has resurfaced again more recently with the concepts of "structure–activity landscapes"[64,65] and "bioactivity-guided navigation of chemical space" using clustering tools in protein space Protein Structure Similarity Clustering (PSSC) and biology-oriented synthesis (BIOS)[66]. We wish to discuss the similarities and differences between these concepts and the AtlasCBS scheme proposed here to explore ways in which each concept can enrich the other for the maximum benefit of the drug-discovery community.

A key difference between AtlasCBS and ChemGPS has been mentioned above. The latter does not directly include any affinity data in the parameter set used to describe the chemical entities. However, it does include drug and nondrug molecules as satellites to guide the navigation but the affinity of ligands is not consulted during the navigation process. The analysis of the physicochemical properties of the molecules is all inclusive and the coordinates are t-scores extracted by principal component analysis (PCA) from as many as 72 descriptors to a total set of 423 satellites and reference structures. The space is considered to be n-dimensional based on the results of the PCA and, if certain planes provide robust description, projection methods can be used. All of these features represent a sound foundation to explore chemical space on its own that could possibly be used in AtlasCBS in the future.

In contrast to ChemGPS, the more recent concepts related to "structure–activity landscapes" do use the affinity (activity) data as an additional variable (typically a third dimension (z)), related to chemical descriptors in the x,y plane(s). Indeed, the geographical and navigational metaphors are obvious when talking about "activity landscapes" or recently "activity cliffs"[64, 65]. The term "navigation" has also been used to guide the synthetic efforts (BIOS) in chemistry space, focused on extending scaffolds found among the chemical subspace of natural products (structural classification of natural products [SCONP]), using protein homologies in the active sites of various enzymes[66]. Those similarities aside, the concept proposed in AtlasCBS is different in various respects.

First, analysis of the SAR landscapes is still based only on the potency as the dominant variable; no account is made for factoring in the size or polarity of the molecules. Second, the descriptors on the x,y directions are similarity descriptors (i.e., Tanimoto scores) in relation to other chemical entities and not the individual molecules *per se*. Thirdly, the overall objective of these most recent analyses of CBS is not to produce a comprehensive map (or atlas) but rather to analyze the individual properties of the activity maps locally and globally to make sound inferences from the available SAR data. Further relationships between the concepts of "activity cliffs", "activity ridges," and other ways to describe the SAR data in drug discovery will be discussed in Part II, when the idea of "trajectories" derived from the AtlasCBS framework are introduced. Currently, there is not enough data to assess the impact of "cliffs" or "ridges" on the overall description of CBS. Perhaps in the future, some of these thoughts could help to refine certain variables currently incorporated into the AtlasCBS formulation or to incorporate new ones.

One last note should be added in regard to the chemical subspace of natural products. Examples will be shown later of how the complexity of natural products (e.g., bioactive peptides) can still be mapped in the proposed AtlasCBS framework. This issue requires further exploration. The natural product community is encouraged to explore the framework proposed here to incorporate and study this critical subset of CBS and trying to relate the possible natural product "regions" with the simpler "drug-like" territories as explored by the work of Bon and Waldmann[66].

3.5 DISCUSSION

Although the AtlasCBS concept is an attractive idea, it is only a first attempt to use combined variables, beyond affinity alone, to map CBS and to try to extract novel insights from the immense amounts of data available in SAR databases. Additional variables that would expand the chemical description of the ligands and their relation to the biological targets will be eventually needed. Within the graphical framework discussed, it should be possible to expand the NPOL descriptor from the current 1D parameter to a full plane that would break down the character of molecules containing N and O and possibly the different kinds of chemical environments for each of these two distinct atoms. It is likely that more variables (besides size and polarity) will be needed to enhance the predictive value and to facilitate and expedite drug discovery in a robust manner. However, maintaining a unified and consistent definition of a limited set of variables will add to its simplicity and utility. The question of the minimum set of variables required to have the most predictive power is still open. Extensive use and testing by the community is encouraged to assess its utility as a mapping tool and to test its robustness as a prospective guide for drug discovery.

CHAPTER 4

The AtlasCBS Application

The concepts summarized and illustrated in the preceding sections and in previous publications[40,48,49] introduced the concepts of an atlas-like representation of chemicobiological space (CBS) to guide us in the search for those CBS regions where the best preclinical candidates might be located. Can we possibly put all these notions and ideas into a software tool that could (i) facilitate their wider use for the drug-discovery community and (ii) test their robustness as a predictor of drug-discovery efforts? When the original ideas were put forward a few years ago, the possibility appeared to be remote. Grant applications did not receive good reviews. Fortunately, the support of colleagues from the Spanish scientific community (Prof. F. Gago, University of Alcalá de Henares and Dr. Antonio Morreale, Centro de Biología Molecular Severo Ochoa, both institutions in the Madrid area) and the Spanish government made it possible to develop the AtlasCBS server that has recently been made available to the global drug-discovery community. The article describing this Internet tool was published recently[67] and will be used as a key reference to describe the main elements and features of the server. Further technical details can be found in the published paper and on the web site (https://www.ebi.ac.uk/chembl/atlascbs). This section could serve as an initial user guide, but there is no substitute for using the software and exploring its power and limitations. The community is encouraged to use the resource and assess its value for the individual projects both in academia and in the private sector.

4.1 BASIC ELEMENTS

The main web page for the server contains the five tabs shown in Fig. 4.1 (upper left). Briefly, they can be described as follows.

Main. Containing basic information about the server and its purpose, main references, contact information, and access to the main topics covered in the **Help** tab.

Map viewer. This is the core of the graphics engine of the server. It contains tools for uploading the data from existing databases and for visualizing and analyzing their content.

Login. Required if the user wants to have private database access. This tab changes to **Manage data** upon a successful proprietary login. This makes available to the user a "UserSet" containing the personal data.

Help. Basic information on how to use the AtlasCBS server.

About. Contains information concerning the institutions and people involved in the AtlasCBS project.

These five tabs in the portal contain the core modules that provide all the functionalities for the server.

4.2 FUNCTIONALITIES

Map viewer. This is the central element of the AtlasCBS server. The graphical engine of the server represents data from different sources, allows visualization of the chemical structure and provides efficient filtering tools to compare and classify the molecules. The users can upload, select, and map any data source target available at BindingDB, PDBBind, ChEMBL, or load an external data set (Module 2). The application opens in the **Data** tab within the **Map Viewer** panel. The user should select the data source by selecting target and organism. A lower window indicates the number of affinity data available for K_i, IC_{50}, and K_d. Selecting the **Add source** tab makes the existing data available to the **Map viewer** and an "efficiency plane" containing the corresponding target—ligand points are represented in a default color on the map. By default the variables shown are NSEI (x) and nBEI (y). This selection is justified because these maps are very easy to interpret. The data points are clustered along lines of slope NPOL

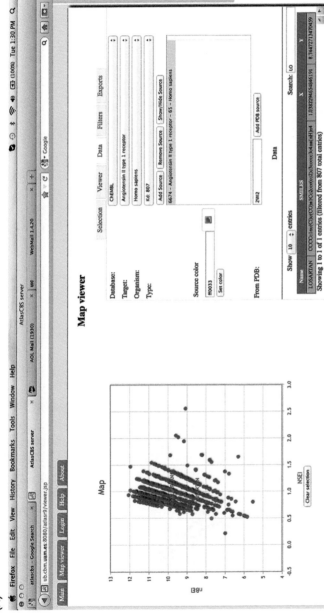

Figure. 4.1 *Example screens presented by the AtlasCBS server. (A) The upper left tabs correspond to the main pages of the server: **Main**, **Map viewer**, **Login** (for private access), **Help**, and **About** (see text). The left graphical panel within the page represents a typical efficiency plane (nBEI vs. NSEI) for the 807 entries found in ChEMBL for angiotensin II type I receptor with K_d affinity values (see Type window). Each point in the plane represents a target–ligand pair. The angular coordinate (NPOL in this case) corresponds to the number of polar (N,O) atoms of the ligand increasing counterclockwise (NPOL=3–12). The radial coordinate corresponds to the affinity of the ligands toward the target. The top right panel shows the different options for the management of the session within the **Map viewer** and choice of database, target, and organism. Most importantly, the choice of LEIs as Cartesian axes (x,y) (within **Viewer**) that determines the appearance of the efficiency planes. The lower right panel shows the compounds selected listed in alphabetic order. The values of the LEIs variables are shown also in this window along with the SMILES strings of the corresponding compounds. The SMILES string of one of the selected compounds (Losartan) is shown in the lower right-hand panel, resulting from inputting "LO" in the search box. It is also annotated on the left map with black lettering. Using the "Duplicate window" tab, multiple windows can be displayed simultaneously to compare pages in the "AtlasCBS" with different variables or scales, as in a real-life atlas.*

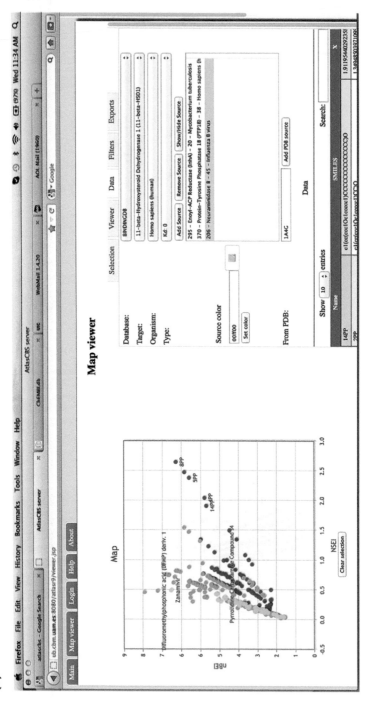

Figure. 4.1 (B) A different screen presented by the AtlasCBS server. In this case, the map in the NSEI–nBEI plane corresponds to the response of the server to three successive PDB access codes: 2H7L: Enoyl-ACPReductase (InhA) from Mycobacterium tuberculosis (red); 2CNE: human protein tyrosine phosphatase 1B (amber); and 144G, Neuraminidase B (green). The colors were changed for contrast using the "Source color" option. In addition to the compound markers corresponding to the three PDB entries, other compounds have been highlighted (black lettering 14 PP, 7 PP, etc.). Note that the marketed drug zanamivir occupies the top (most efficient) position of the Neuraminidase B inhibitors (green set) (see Part III). Reprinted and adapted from Cortes-Cabrera et al.[67] Copyright J. Comp. Aided Mol Design.

(N + O count), and increasing counterclockwise from low polarity (low NPOL) to more polar ligands (higher values of NPOL). As the NPOL values increase, the lines are closer together near the vertical direction and a change in scale might be needed to clearly distinguish the trends.

This initial efficiency plane can be changed by the appropriate selection of axes (x,y) in the **Viewer** tab. A brief definition of the possible variables in tabular form is presented by opening a pull-down tab and further details are provided in links. The existing variables are (clustered in the suggested x,y pairs): SEI, BEI; NSEI, NBEI; NSEI with nBEI or mBEI, respectively. Any combination is possible but combinations of size/polarity (x/y) are strongly suggested to maintain a consistent interpretation of the different efficiency planes or maps.

Given a map, molecules can then be selected by clicking on the corresponding point in the map. A brief description, taken from the existing data in the databases, is shown on the screen just below the corresponding target–ligand point. The 2D structure of the selected molecule can be seen in full detail within the **Selection** tab, including its basic physicochemical properties. Depending upon the origin of the data, a direct link to the corresponding database (i.e., BindingDB) would also appear. A user can also select molecules by typing a few letters in the search window: for example, typing "LO" will result in "LOSARTAN" being selected. A window in the lower right panel shows the selected compounds in alphabetic order. The list contains compound name, SMILES strings, and the corresponding ligand efficiency variables. These selected compounds are also annotated on the map in black lettering. The color of the points related to the "target" set can be changed for contrast using the *Source color* option and it is extremely useful to do so when displaying several targets (Fig. 4.1B). Selecting a certain subset of the efficiency plane with the mouse will result in the display of the corresponding subset with the axes rescaled. In addition, by using the **Duplicate window** tab, multiple windows can be displayed simultaneously with different variables and scales giving the feeling of examining multiple pages of an atlas simultaneously.

Depending on the browser used to run the applications, other selection tools are available within the **Filter tab** using different ranges of values or certain slopes in the NSEI–nBEI planes. Finally, one can save and restore any working session and export images and data. Further details are given in the original publication[67].

Besides the direct access to the available data in BindingDB, ChEMBL, or PDBBind, another way to access and upload target−ligand data is via the protein data bank (PDB) accession code of protein−ligand complexes. This provides a convenient link to structural and binding data. Within **Data**, the user can input directly a PDB access code and the server will plot in the default NSEI−nBEI efficiency plane all the affinity data present in BindingDB for that target. Typically, the affinity parameter (IC$_{50}$, K_i, or K_d) with the larger number of entries is selected by default. An unsuccessful search will prompt the user for a different access code based on further exploration of the content of BindingDB for that target. For these purposes, the direct link between the AtlasCBS server and BindingDB is critical as not all searches are successful. Examples of the versatility and results of this option are shown in Figs. 4.1B and 4.2B.

The main purpose of the AtlasCBS server is to allow the users to examine and explore the effectiveness of the underlying concepts of ligand efficiency (LE) in their drug-discovery projects. To this end, the server allows users to upload and process their own datasets in a secure way. A simple registering process based on a valid e-mail address and user-chosen password opens the access to the AtlasCBS for proprietary use. Upon a successful login, the **Manage data** tab appears and the **UserSet** is added to the list of the available databases. Two types of private datasets are permitted currently for further study and comparison with existing databases:

1. Conventional affinity data from any project can be uploaded from an external file (semicolon-separated values, equivalent to a Comma Separated Value [CSV], format) containing the following variables: molecule name, SMILES code, type of affinity/activity variable (K_i, IC$_{50}$, K_d), and the corresponding value in nanomolar concentration units.

2. The content of chemical libraries can also be uploaded into the private data stream (without any affinity data). The files containing only the compound name/number within the library (e.g., AS0045) and the corresponding SMILES string can be uploaded. The server assigns random affinity K_i values in the micro-nanomolar range and plots them on the default NSEI−nBEI plane. To understand these plots and their implications in terms of the composition of the chemical library, we refer the reader to an earlier publication[49] and

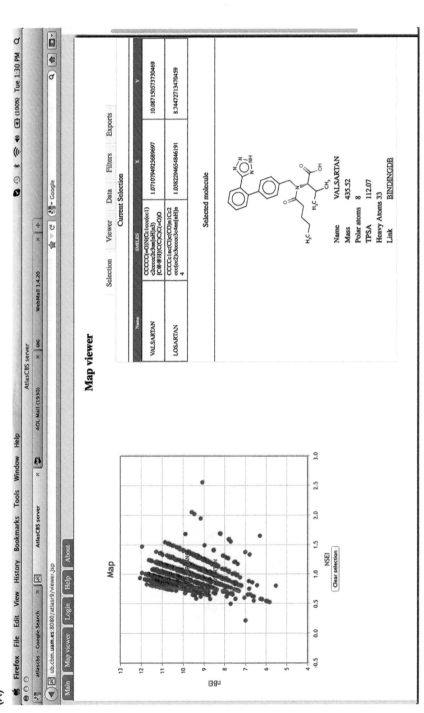

Figure. 4.2 Mapping efficiency and structure in CBS. (A) Image from the **Map Viewer** panel illustrating the position of a certain selected compound targeting angiontensin II type 1 receptor (as in Fig. 4.1A) (Valsartan) and the corresponding 2D structure within the **Selection** tab. The basic physicochemical parameters of the compound are displayed as well as a direct link to BindingDB that can be used to look up some additional information. The relative efficiency of other compounds binding to the same target such as Losartan (Fig. 4.1A) can be compared.

(B)

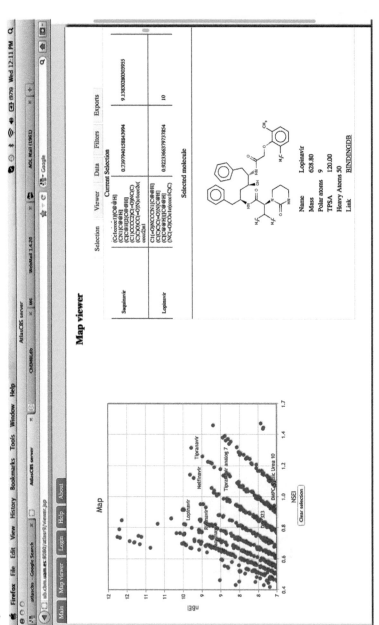

Figure. 4.2 (B) Close-up image from the **Map viewer** panel showing a subset of the compounds extracted from BindingDB as a response to the PDB access code 1OHR (Nelfinavir). The 2D structure of the selected compound is shown. The progressive migration toward higher efficiencies of the compounds targeting the HIV-1 protease can be seen. The structure of Lopinavir, a second generation HIV-1 protease inhibitor in the market, is shown for reference. The angular component (slope of the lines) relates to the number of polar atoms $(N + O)$. The radial coordinates are given by the measured affinity toward the corresponding target. Different measurements for different targets (i.e., wild type vs. mutant(s)) will all map along the same line. The thickness of the lines depends on the number of compounds with the same NPOL number but different number of NHA atoms ($\log_{10}(NHA)$, see Table 2.1. The selection of this area (NSEI: 0.4–1.7; nBEI: 7–12) was made by highlighting this region of the full NSEI–nBEI plane with the mouse upon the representation of all the compounds on the plane for the HIV protease target. The physicochemical properties of the ligand are shown with a direct link to BindingDB as above. Note how Lopinavir, Tipranavir, Nelfinavir occupy the highest efficiency positions for the corresponding NPOL lines in relation to other analogs (i.e., Tipranavir analog 7, third line from the right is below the final compound). Reprinted and adapted from Cortes-Cabrera et al.[67] Copyright J. Comput. Aided Mol Design.

the following section of this book (Part II), where conceptual applications are discussed.

As noted before, to generate a map, the user can select x and y variables from the sets: SEI, BEI; NSEI, NBEI; NSEI, nBEI, or NSEI with mBEI. Any combination is possible, but complementary pairs (ligand efficiency indices [LEIs] per size and polarity) are recommended.

Data from the existing public databases included in the server and proprietary data extracted from the **UserSet** can be displayed simultaneously allowing for an extremely useful direct comparison. However, it should be noted that the same chemical entity tested under different assay conditions would yield different affinity values. This would result in the mapping of the same chemical compound along the same "chemical line" but at different radial distances from the origin (different biological assays). Any rigorous comparison of the different efficiencies of various compounds from different sources should be based on affinity values obtained under the same assay conditions and comparing the same affinity measurement. Comparison of the LEI calculated from different affinity variables (K_i vs. IC_{50} or K_d) should be done with prudence.

A brief description of the implementation details of the server is included here to better understand the possibilities, limitations, and further enhancements of the AtlasCBS. The server is organized in three layers: clients, the application server, and an internal database. Each layer can communicate with the nearest neighbor but not beyond. The three layers have been implemented with the following elements for the different components: (i) Java, JavaScript, and HTML clients; (ii) the Apache Tomcat servlet container; and (iii) the MySQL database engine.

The information contained in the current releases of BindingDB (19/05/2012), PDBBind (v2011), and ChEMBL (v13) was imported into the MySQL server's database by stand-alone Java programs that also computed the molecular properties and efficiency indices as defined and described previously (Table 2.1 and other publications). The molecular properties (MW, NPOL, NHA, and PSA) are calculated using the Chemistry Development Toolkit (CDK)[67] within the Java external programs. Only the three best characterized affinity parameters (K_i, IC_{50}, K_d) are extracted from the myriad of data existing in the SAR databases discussed previously. User-uploaded external data are processed "on the fly" with a special servlet also using the CDK. Thus, the content of the AtlasCBS server is dependent on the most recent data used to create the corresponding version. We acknowledge that this is a limitation, but the users could create their own datasets on specific projects, targets or even chemical series, and upload the data to map the quality of their compounds in relation to the existing chemical entities available for that target. This has the advantage of having the affinity data for different compounds (including competing series) obtained from a more "controlled" set of conditions for the assays, coming from the same laboratory and therefore making relative comparison compounds more reliable.

4.3 DISCUSSION

The relative merits as well as the equivalences of the different LEI and size-related definitions have already been discussed. The advantages for the proposed numerical and analytical framework have been presented. Namely, this formulation provides a simple representation of the data in SAR databases and provides a unified numerical framework that combines three essential parameters for drug discovery (affinity, size, polarity) into an attractive 2D representation. Nonetheless, this software tool represents only the first attempt to translate the AtlasCBS concept into a practical tool. There are probably many other ways to design and execute the concept.

In addition, the current version is far from encompassing the broad range of variables and possibilities outlined in the previous chapter. The team that created the AtlasCBS is already planning to add other LEIs into the framework to facilitate the critical assessment of the different metrics both individually and in combination. In particular, the inclusion of the original definition by Hopkins et al.[18] (Eq. (2.1), Table 2.2), as it is widely used[33]. Size-corrected indices and entropic and enthalpic ligand efficiencies are also being considered for inclusion into the framework. The ability to display trajectories and estimate "distances" from two points in CBS is also being studied as a future enhancement. Input from the user community is important to continue to determine the future of this resource. The notions and concepts put forward need to be used, explored, and tested if they are to provide a robust foundation for effective drug discovery in the future.

REFERENCES FOR PART I

1. Berman HM. The Protein Data Bank: a historical perspective. Acta crystallographica Section A, Foundations of crystallography. 2008; **64**(Pt 1): 88−95.

2. Irwin JJ, Shoichet BK. ZINC—a free database of commercially available compounds for virtual screening. Journal of chemical information and modeling. 2005; **45**(1): 177−82.

3. Bairoch A, Apweiler R. The SWISS-PROT protein sequence database and its supplement TrEMBL in 2000. Nucleic acids research. 2000; **28**(1): 45−8.

4. Kramer C, Kalliokoski T, Gedeck P, Vulpetti A. The experimental uncertainty of heterogeneous public K(i) data. Journal of medicinal chemistry. 2012; **55**(11): 5165−73.

5. Southan C, Varkonyi P, Muresan S. Quantitative assessment of the expanding complementarity between public and commercial databases of bioactive compounds. Journal of cheminformatics. 2009; **1**(1): 10.

6. Bellis LJ, Akhtar R, Al-Lazikani B, Atkinson F, Bento AP, Chambers J, et al. Collation and data-mining of literature bioactivity data for drug discovery. Biochemical society transactions. 2011; **39**(5): 1365−70.

7. Degtyarenko K, de Matos P, Ennis M, Hastings J, Zbinden M, McNaught A, et al. ChEBI: a database and ontology for chemical entities of biological interest. Nucleic acids research. 2008; **36**(Database issue): D344−50.

8. Abad-Zapatero C. Notes of a protein crystallographer: on the high-resolution structure of the PDB growth rate. Acta crystallographica Section D, Biological crystallography. 2012; **68** (Pt 5): 613−7.

9. Berman H, Henrick K, Nakamura H. Announcing the worldwide Protein Data Bank. Nature structural biology. 2003; **10**(12): 980.

10. Lawson CL, Baker ML, Best C, Bi C, Dougherty M, Feng P, et al. EMDataBank.org: unified data resource for CryoEM. Nucleic acids research. 2011; **39**(Database issue): D456−64.

11. Wang R, Fang X, Lu Y, Wang S. The PDBbind database: collection of binding affinities for protein−ligand complexes with known three-dimensional structures. Journal of medicinal chemistry. 2004; **47**(12): 2977−80.

12. Hu L, Benson ML, Smith RD, Lerner MG, Carlson HA. Binding MOAD (Mother Of All Databases). Proteins. 2005; **60**(3): 333−40.

13. Gaulton A, Bellis LJ, Bento AP, Chambers J, Davies M, Hersey A, et al. ChEMBL: a large-scale bioactivity database for drug discovery. Nucleic acids research. 2012; **40**(Database issue): D1100−7.

14. Olah M, Rad R, Ostopovici L, Bora A, Hadaruga N, Hadaruga D, et al. WOMBAT and WOMBAT-PK: bioactivity databases for lead and drug discovery. In: Schreiber SL, Kapoor T, Wess G, editors. Chemical Biology: From Small Molecules to Systems Biology and Drug Design. Weinheim: Wiley-VCH Verlag CmbH & Co.; 2007.

15. Weininger D. SMILES 1: Introduction and encoding rules. Journal of chemical information and computer science. 1988; **28**: 31−6.

16. Weininger D, Weininger A, Weininger JL. SMILES 2. Algorithm for generation of unique SMILES notation. Journal of chemical information and computer science. 1989; **29**: 97−191.

17. Lipinski CA, Lombardo F, Dominy BW, Feeney PJ. Experimental and computational approaches to estimate solubiliy and permeability in drug discovery and development settings. Advances drug delivery reviews. 1997; **23**: 3−25.

18. Hopkins AL, Groom CR, Alex A. Ligand efficiency: a useful metric for lead selection. Drug discovery today. 2004; **9**(May): 430−1.

19. Wishart DS, Knox C, Guo AC, Shrivastava S, Hassanali M, Stothard P, et al. DrugBank: a comprehensive resource for in silico drug discovery and exploration. Nucleic acids research. 2006; **34**(Database issue): D668−72.

20. Chen X, Liu M, Gilson MK. BindingDB: a web-accessible molecular recognition database. Combinatorial chemistry and high throughput screening. 2001; **4**(8): 719−25.

21. Chen X, Lin Y, Gilson MK. The binding database: overview and user's guide. Biopolymers. 2001; **61**(2): 127−41.

22. Chen X, Lin Y, Liu M, Gilson MK. The binding database: data management and interface design. Bioinformatics. 2002; **18**(1): 130−9.

23. Bolton EE, Wang Y, Thiessen PA, Bryant SH. PubChem: Integrated Platform of Small Molecules and Biological Activities. Annual Reports in Computational Chemistry. 2008; **4**: 217−41.

24. Li Q, Cheng T, Wang Y, Bryant SH. PubChem as a public resource for drug discovery. Drug discovery today. 2010; **15**(23−24): 1052−7.

25. Nicola G, Liu T, Gilson MK. Public domain databases for medicinal chemistry. Journal of medicinal chemistry. 2012; **55**(16): 6987−7002.

26. Abad-Zapatero C. A sorcerer's apprentice and the Rule of Five: from rule of thumb to commandments and beyond. Drug discovery today. 2007; **12**(23−24): 995−7.

27. Knox C, Law V, Jewison T, Liu P, Ly S, Frolkis A, et al. DrugBank 3.0: a comprehensive resource for "omics" research on drugs. Nucleic acids research. 2011; **39**(Database issue): D1035−41.

28. Liu T, Lin Y, Wen X, Jorissen RN, Gilson MK. BindingDB: a web-accessible database of experimentally determined protein−ligand binding affinities. Nucleic acids research. 2007; **35**(Database issue): D198−201.

29. Andrews PR, Craik DJ, Martin JL. Functional group contributions to drug−receptor interactions. Journal of medicinal chemistry. 1984; **27**(12): 1648−57.

30. Kuntz ID, Chen K, Sharp KA, Kollman PA. The maximal affinity of ligands. Proceedings of the national academy of sciences of the United States of America. 1999; **96**(18): 9997−10002.

31. Pelphrey PM, Popov VM, Joska TM, Beierlein JM, Bolstad ES, Fillingham YA, et al. Highly efficient ligands for dihydrofolate reductase from *Cryptosporidium hominis* and *Toxoplasma gondii* inspired by structural analysis. Journal of medicinal chemistry. 2007; **50**(5): 940−50.

32. Saxty G, Woodhead SJ, Berdini V, Davies TG, Verdonk ML, Wyatt PG, et al. Identification of inhibitors of protein kinase B using fragment-based lead discovery. Journal of medicinal chemistry. 2007; **50**(10): 2293−6.

33. Ligand efficiency poll results. 2011. Available from: <http://practicalfragments.blogspot.com/2011/08/ligand-efficiency-metrics-poll-results.html>.

34. Freire E. Do enthalpy and entropy distinguish first in class from best in class? Drug discovery today. 2008; **13**(19−20): 869−74.

35. Freire E. A thermodynamic approach to the affinity optimization of drug candidates. Chemical biology and drug design. 2009; **74**(5): 468−72.

36. Ferenczy GG, Keseru GM. Enthalpic efficiency of ligand binding. Journal of chemical information and modeling. 2010; **50**(9): 1536−41.

37. Reynolds CH, Holloway KM. Thermodynamics of ligand binding and efficiency. ACS medicinal chemistry letters. 2011; **2**: 433−7.

38. Garcia-Sosa AT, Hetenyi C, Maran U. Drug efficiency indices for improvement of molecular docking scoring functions. Journal of computational chemistry. 2009; **31**(1): 174–84.

39. Verdonk ML, Rees DC. Group efficiency: a guideline for hits-to-leads chemistry. ChemMedChem. 2008; **3**(8): 1179–80.

40. Abad-Zapatero C, Metz JM. Ligand efficiency indices as guideposts for drug discovery. Drug discovery today. 2005; **10**(7): 464–9.

41. Abad-Zapatero C. Ligand efficiency indices for effective drug discovery. Expert opinion in drug discovery. 2007; **2**(4): 469–88.

42. Abad-Zapatero C, Perisic O, Wass J, Bento PA, Overington J, Al-Asikani B, et al. Ligand efficiency indices for an effective mapping of chemico-biological space: the concept of an atlas-like representation. Drug discovery today. 2010; **15**(19–20): 804–11.

43. Gracias V, Ji Z, Akritopoulou-Zanze I, Abad-Zapatero C, Huth JR, Song D, et al. Scaffold oriented synthesis. Part 2: Design, synthesis and biological evaluation of pyrimido-diazepines as receptor tyrosine kinase inhibitors. Bioorganic and medicinal chemistry letters. 2008; **18**(8): 2691–5.

44. Lepifre F, Christmann-Franck S, Roche D, Leriche C, Carniato D, Charon C, et al. Discovery and structure-guided drug design of inhibitors of 11beta-hydroxysteroid-dehydrogenase type I based on a spiro-carboxamide scaffold. Bioorganic and medicinal chemistry letters. 2009; **19**(13): 3682–5.

45. Roche D, Carniato D, Leriche C, Lepifre F, Christmann-Franck S, Graedler U, et al. Discovery and structure–activity relationships of pentanedioic acid diamides as potent inhibitors of 11beta-hydroxysteroid dehydrogenase type I. Bioorganic and medicinal chemistry letters. 2009; **19**(10): 2674–8.

46. Hajduk PJ. Fragment-based drug design: how big is too big? Journal of medical chemistry. 2006; **49**: 6972–6.

47. Tanaka D, Tsuda Y, Shiyama T, Nishimura T, Chiyo N, Tominaga Y, et al. A practical use of ligand efficiency indices out of the fragment-based approach: ligand efficiency-guided lead identification of soluble epoxide hydrolase inhibitors. Journal of medicinal chemistry. 2011; **54**(3): 851–7.

48. Abad-Zapatero C, Perisic O, Wass J, Bento AP, Overington J, Al-Lazikani B, et al. Ligand efficiency indices for an effective mapping of chemico-biological space: the concept of an atlas-like representation. Drug discovery today. 2010; **15**(19–20): 804–11.

49. Abad-Zapatero C, Blasi D. Ligand efficiency indices (LEIs): more than a simple efficiency yardstick. Molecular informatics. 2011; **30**(2–3): 122–32.

50. Lipinski CA. Drug-like properties and the causes of poor solubility and poor permeability. Journal of pharmacological and toxicological methods. 2000; **44**(1): 235–49.

51. Reynolds CH, Tounge BA, Bembenek SD. Ligand binding efficiency: trends, physical basis, and implications. Journal of medicinal chemistry. 2008; **51**(8): 2432–8.

52. Willem J, Nissink JW. Simple size-independent measure of ligand efficiency. Journal of chemical information and modeling. 2009; **49**(6): 1617–22.

53. Orita M, Ohno K, Niimi T. Two "Golden Ratio" indices in fragment-based drug discovery. Drug discovery today. 2009; **14**(5–6): 321–8.

54. Leeson PD, Springthorpe B. The influence of drug-like concepts on decision-making in medicinal chemistry. Nature reviews drug discovery. 2007; **6**(11): 881–90.

55. Mortenson PN, Murray CW. Assessing the lipophilicity of fragments and early hits. Journal of computer-aided molecular design. 2011; **25**(7): 663–7.

56. Keseru GM, Makara GM. The influence of lead discovery strategies on the properties of drug candidates. Nature reviews Drug discovery. 2009; **8**(3): 203–12.

57. Ladbury JE, Klebe G, Freire E. Adding calorimetric data to decision making in lead discovery: a hot tip. Nature reviews drug discovery. 2010; **9**(1): 23–7.

58. Wager TT, Hou X, Verhoest PR, Villalobos A. Moving beyond rules: the development of a central nervous system multiparameter optimization (CNS MPO) approach to enable alignment of druglike properties. ACS chemical neuroscience. 2010; **1**(6): 435–49.

59. Braggio S, Montanari D, Rossi T, Ratti E. Drug efficiency: a new concept to guide lead optimization programs towards the selection of better clinical candidates. Expert opinion on drug discovery. 2010; **5**(7): 609–18.

60. Bickerton GR, Paolini GV, Besnard J, Muresan S, Hopkins AL. Quantifying the chemical beauty of drugs. Nature chemistry. 2012; **4**(2): 90–8.

61. Hann MM, Keseru GM. Finding the sweet spot: the role of nature and nurture in medicinal chemistry. Nature reviews drug discovery. 2012; **11**(5): 355–65.

62. Lipinski C, Hopkins A. Navigating chemical space for biology and medicine. Nature. 2004; **432**: 855–61.

63. Oprea TI, Gottfries J. Chemography: the art of navigating in chemical space. Journal of combinatorial chemistry. 2001; **3**(2): 157–66.

64. Bajorath J. Modeling of activity landscapes for drug discovery. Expert opinion on drug discovery. 2012; **7**(6): 463–73.

65. Bajorath J, Peltason L, Wawer M, Guha R, Lajiness MS, Van Drie JH. Navigating structure–activity landscapes. Drug discovery today. 2009; **14**(13–14): 698–705.

66. Bon RS, Waldmann H. Bioactivity-guided navigation of chemical space. Accounts of chemical research. 2010; **43**(8): 1103–14.

67. Cortes-Cabrera A, Morreale A, Gago F, Abad-Zapatero C. AtlasCBS: a web server to map and explore chemico-biological space. Journal of computer-aided molecular design. 2012; **26**(9): 995–1003.

PART II

Conceptual Applications of the AtlasCBS: A New Paradigm

Applications of the AtlasCBS concept are presented to clarify the initial description and to illustrate how it can be used in various areas of drug discovery. Examples are presented in three general areas, one per chapter. Chapter 5: Analysis of the contents of SAR databases: (i) Graphical representation of Lipinski's Rule of Five; (ii) The territory of drugs vs. nondrugs; and (iii) Mapping of bioactive peptides. Chapter 6: Fragment-based strategies: (i) Analysis of fragment or chemical libraries, and (ii) Fragment deconstruction. Chapter 7: Navigating in chemicobiological space: (i) Trajectories in chemicobiological space, and (ii) The search for optimal trajectories. Some of these conceptual applications will be developed further in Part III and illustrated with specific examples using the AtlasCBS Internet application.

PERSONAL INTRODUCTION

After the first ideas expanding the definitions of LEIs were published in 2005, I needed to apply the concepts to specific examples to convince my colleagues and myself that these early steps could have an application to expedite drug design. Some medicinal chemists began to use the size-related efficiency index to monitor their projects and the increasing interest in fragment-based methodologies and strategies aided in the dissemination and application of these ideas[1]. My interest was to map the drug-discovery process on a SEI–BEI plane. I applied the early LEIs (SEI, BEI) to a couple of projects in which I had been directly involved and included the analysis within an overview of where I thought LEIs could take us in the future[2]. In this publication, an attempt was made to map most of the compounds designed at Abbott Laboratories against human protein tyrosine phosphatase 1B (hPTP1B) on the SEI–BEI plane; it was very revealing. I added a few drugs (i.e., Haloperidol, Iressa) and other reference compounds onto the map, and the plot suggested that the vast variety of chemical substances could be enclosed into a wedge bounded by the PSA/MW ratios of the compounds from the most hydrophobic to the highly polar[2]. Could this be true? The compounds for the different series of inhibitors for hPTP1B were still all clustered together. Were there other variables that could facilitate the interpretation?

Then came a professional transition that permitted me to dedicate more time to these ideas. During the summer of 2007, Prof. Michael Johnson, director of the Center for Pharmaceutical Biotechnology (CPB) at the University of Illinois at Chicago (UIC) offered me a programmer for the summer. I saw my opportunity. Dr. Ognjen Perišić and I worked for several months trying to map the full content of several SAR databases onto a series of efficiency planes. First, we tried PDBBind using SEI–BEI and later the indices based on atomic (atom counting) properties: NSEI, NBEI. During some of those algebraic manipulations, I discovered the simplicity and clarity of the nBEI vs. NSEI plane. In this exploration, my colleague from Abbott Laboratories, John Wass, assisted me in the statistical analysis. From then on, I thought that we could use these variables to map CBS. Those ideas were materialized conceptually during a 2-month stay in the fall of 2009 at ChEMBL hosted by Dr. John Overington and Dr. Janet Thornton (director of EBI). It was during those 2 months that working with Dr. Patricia Ana Bento and the ChEMBL group, using Spotfire™ and other software tools, that the tangible notion of an atlas of CBS became a reality and the first maps were produced[3].

In parallel, some medicinal chemists were using consistently BEI to monitor their progress within series and across different series[1,4,5]. My immediate response to those publications was: Why don't they use

BEI *and* SEI? I did contact some of the computational chemists in those groups and I proposed to them the use of both indices as a superior predictive strategy. A collaboration was arranged and this is how the notion of "trajectories" in efficiency planes originated. These were the three initial threads of thought that gave birth to the AtlasCBS concept. What follows will illustrate these ideas with concrete examples. Not all these conceptual applications are fully implemented yet in the current AtlasCBS server, but I feel that reviewing these ideas will help to have a perspective of the past and make us think into the future.

Analysis of the Content of SAR Databases

As documented before, the data relating the biological targets to the chemical substances that modulate their action is enormous and will continue to grow as new technologies and additional biomedical knowledge continues to accumulate. Therefore, we need to find ways to represent and chart all that information in ways that can help us to get further insights into the relationship between targets and ligands and their relative importance in the drug discovery process. Can we extract some general principles, somewhat equivalent to the Kepler's laws of planetary motion, that would help us to develop a comprehensive description of the target–ligand universe? In this chapter, three applications are presented to show how effective the AtlasCBS representation is in mapping the overall content of SAR databases, from small molecules to large peptides. More specific examples will be presented in Part III. I surmise that other applications and insights will come in the near future.

5.1 GRAPHICAL REPRESENTATION OF LIPINSKI'S GUIDELINES (Ro5)

Lipinski's Rule of Five (Ro5) has probably been the most influential concept in preclinical drug discovery during the last decade.[7] The notion was proposed by Chris Lipinski and coworkers in 1997 in an attempt to extract some guidelines or general notions as to the properties of molecules that were likely to be not permeable or poorly absorbed. It was formulated as an easy mnemonic related to number five. In essence, it stated that poor passive absorption or permeability of a compound is more likely if the compound violate two or more of the following conditions; if they have: (i) >5 hydrogen-bond donors; (ii) molecular mass >500; (iii) calculated $\log P > 5$; and the sum of donors and acceptors $(N + O) > 10$.[6] Alternatives to the Ro5 using other molecular descriptors such as polar surface area (PSA) or other parameters related to the surface/volume of the molecules were also proposed soon thereafter.[7,8] Extensions and variations of these guidelines appeared later when the number of rotatable bonds and molecular flexibility were also recognized to be factors important for bioavailability.[9] A "Rule of Three" (Ro3) adaptation for the molecular properties of fragments was proposed by a group at Astex.[10]

Because of its simplicity and ease of implementation in the drug discovery computational protocols, particularly at the pruning of high throughput screening results or in the design of chemical libraries, these guidelines were used somewhat uncritically and were soon engraved in the "commandments" of drug discovery as synonyms of "drug-likeness."[11] This extrapolation has had a significant effect on the overall importance of the Ro5 in the drug discovery community, by limiting the exploration of molecules that were not "Ro5 compliant." It should also be noted that the guidelines suggested by the Ro5 did never take into account the affinity of the ligand toward the target, as they were strictly based on the molecular properties of the ligand.

A systematic comparison of the drugs under clinical development with the larger pool of molecules starting Phase I, has indeed shown that there is a consistent decrease in the molecular mass, the number of hydrogen-bond acceptors and to a lesser extent in the number of rotatable bonds.[12] These observations probably influenced the large attrition rates observed in the early combinatorial libraries and have contributed to a more stringent use of the Ro5 guidelines in drug discovery in the past and it is still felt nowadays.

As useful as the Ro5 and other guiding criteria have been, it is fair to say that they only represent a "first approximation" to guide drug discovery. They are restrictive criteria that only have a limited value to help us understand the underlying principles of the target–ligand relationship. Similar views had been expressed before in other publications commemorating the decade of the original publication.[13,14]

An unexpected and surprising insight of the application of LEIs to chart CBS is that they permit a graphical representation of the Ro5 variables of the ligand, in a continuous gradation from the more hydrophobic compounds (low PSA/MW) to the most polar in the angular variable of the efficiency planes (high PSA/MW). This insight is combined with the corresponding information relating the affinity of the chemical entities to the target along the radial component. An example of this representation is shown in Fig. 5.1, where the angular coordinate is color-coded from green to red, counterclockwise. The highest compound along each line corresponds to the most efficient (highest affinity in relation to the size) and the number of polar atoms increases counterclockwise from NPOL $= 1-13$.

Typically, databases (WOMBAT, ChEMBL, and others) contain column entries specifying the number of violations $(1-4)$ of the molecular entity with respect to the Ro5 guidelines. It is possible to use this information and relate it to the physicochemical and atomic (in terms of polar vs. nonpolar) properties of the chemical entities. The polarity of the molecules increases counterclockwise in the diagram and using the number of violations of the Ro5 as a polar coordinate, it is easy to see the regions of plane favored by the Ro5. In the color scheme represented in Fig. 5.2, molecules with $N + O > 10$ will be marked by a sharp boundary for the lines with NPOL > 10. Basically, the angular coordinate (slope) of the lines increases with the first three of Lipinski's guidelines (related nonindependently to polarity of the molecules) and the value of the MW affects the intercept of those lines adding the fourth violation (often the black square). It is obvious then that the graphical view provides a representation of the continuous value of the variables without boundaries and, in addition, adds the critical value of the affinity constant between the ligand and the target to provide a measure of the efficiency of the compounds.

5.2 THE TERRITORY OF DRUGS vs. NONDRUGS

Any explorer would like to have clear indications in the map as to where the treasures are. Drug discoverers should be no different although our riches are different from those sought by fifteenth-century travelers. The initial drug hunters had in mind the original concept of "magic bullets" as a guiding principle: the perfect molecular entity that will be specific for a given clinical condition or infection without affecting any other function. The practices from the past century and our ensuing experiences have shown us otherwise. As visionary as

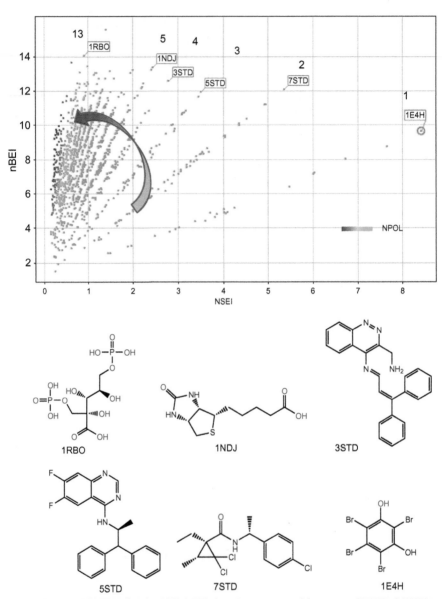

Figure 5.1 *Content of PDBBind in the NSEI–nBEI plane. Representation of the content of PDBBind (2007) in the NSEI–nBEI efficiency plane highlighting the polar description of the variables. The angular coordinate varies from green to red (low to high NPOL numbers) and the most efficient compound for each line is annotated. Compound structures are shown below labeled by the corresponding PDB access code and the polar atoms highlighted (O: red and N: blue).* Reproduced with permission from Ref. [3]. Copyright Drug Discovery Today, 2010. Image produced with Spotfire™.

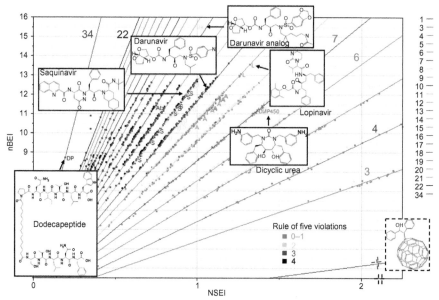

Figure 5.2 NSEI−nBEI efficiency plane for HIV-1 compounds and the Ro5. Representation in the NSEI−nBEI plane of the active compounds against HIV-1 protease available in ChEMBL emphasizing the relationship between the angular coordinate, increasing counterclockwise, and the properties of the molecules related to the Ro5. The number of polar atoms (NPOL) of the compounds has been modeled statistically to produce the corresponding lines indicated on the right panel (NPOL = 1−34). The scale comprises NSEI (0−2.5), nBEI(0−16). There is a sharp boundary when NPOL = 10 with more red/black dots corresponding to the violations of the Ro5. Number of violations is dominated by the polarity (N + O) rules, but compounds with a large molecular mass (intercept of the lines $log_{10}(NHA)$), mapping on the sides of the main line, often add a MW violation (> 500 Da). The fullerene containing one polar atom at the bottom of the scale is off-scale and has been introduced for reference. The analysis was performed with the statistical package JMP9 (SAS Institute). Adapted from Abad-Zapatero and Blassi[16] and reproduced with permission. Copyright Molecular Informatics, 2011.

Erlich's pioneer work with Salvarsan was, it needs to be put in the correct perspective[15] and we (and the public at large) must realize how promiscuous chemical entities are in relation to their biological targets. Even from the very early days of the extensive use of Salvarsan in a broader spectrum of patients, toxicities and morbidities were recognized due to changes in the mode of administration, overdosing, and other factors. Biochemistry does not know the boundaries of the "therapeutical areas" so common in pharmaceutical companies; biochemistry knows about chemical reactions (hydrolysis, condensations, etc.) and knows about enzymatic activities (proteases, kinases, hydrolases to name a few). Thus the demarcations between the affinities of chemical compounds and the various biological targets of therapeutic interests are not very distinct.

The proposed atlas-like representation of CBS can provide an effective way of charting its complexity and show us where the promising territories are to find "drug-like" chemistry in relation to the corresponding targets and in reference to the entire universe of chemical entities. The use of LEIs to represent CBS and illustrate this concept has already been introduced previously and illustrated in the previous section (Figs. 3.2A and B). A different image is presented in Fig. 5.1. The mapping of the compounds included in the PDBBind database (2007 release) in the NSEI (0–9), nBEI (0–16) plane shows a uniformly distributed plane whereby chemical substances were evenly dispersed in the angular coordinate (NPOL = 1–13), within the limits of what is available in the PDB (Fig. 5.1). On the other hand, the large variety of chemical compounds that are active against the HIV-1 protease has been presented in Fig. 5.2, including a wide range of compounds in terms of size and polarity (NPOL = 1–34) in a much narrower range of NSEI (0–2) and similar efficiency values in size (nBEI = 0–16). Can we demarcate regions of CBS where there is a high probability of finding drug-like molecules? Is it possible to establish boundaries in CBS that correspond to specific targets?

An initial mapping with a limited set (~200) of drugs in the market was done by exploring the corresponding data from the early data available from Galapagos NV, prior to its incorporation into the extensive ChEMBL database. This initial result was published earlier and hinted at the concept of an atlas-like representation of CBS.[16] Although limited, the preliminary analysis strongly suggested that the boundaries of the drug-like territory could be expressed in terms of efficiency and polarity. A distinct "exclusion box" in the low size-efficiency variable (nBEI < 6) clearly indicated the importance of binding efficiency, as practically no compounds in the sample were found inside. Alternatively, very few compounds were found with a very high polarity-efficiency (very hydrophobic), corresponding to NSEI > 6. The majority of the sampled drugs were enclosed in a roughly rectangular region with the limits: 1 < NSEI < 5, 6 < nBEI < 12 (Fig. 5.3).

This finding was further explored using the much larger sample of marketed drugs available in WOMBAT (2007) as shown in Fig. 5.4A. The available data include a total of 7,722 affinity measurements (both K_i and IC_{50}) for 1,332 targets. Comparison with the previous image clearly shows that the exclusion region mentioned is also found in this larger

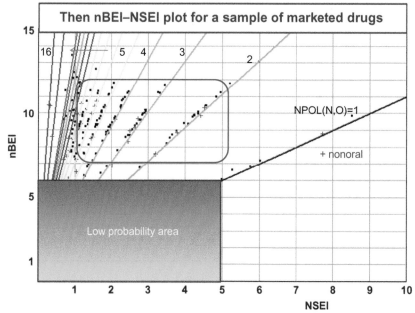

Figure 5.3 Drug territory for a small set. The NSEI–nBEI plane region where a small sample of marketed drugs maps. The yellow box highlights an "exclusion box," where very few molecules lie due to low affinity (too close to the origin on the radial variable). The red rectangle outlines a region where most of the drugs included in the sample are included; note the abundance of orally administered compounds. NPOL increases counterclockwise from blue (very hydrophobic) to red(s) (very polar). + indicates that the compound is administered nonorally. Adapted from Abad-Zapatero and Blasi[16] and reproduced with permission from Molecular Informatics. Copyright Molecular Informatics, 2011.

sample. In addition, a trapezoidal region has been outlined in green where the vast majority of the drugs are enclosed (>90%), corresponding to lines of slope NPOL = 2–15 and the nBEI range of 6 < nBEI < 14.

The versatility and power of the AtlasCBS representation can be appreciated better in Fig. 5.4B where a close-up of the earlier map is also shown. A few features of this more detailed view are worth highlighting. First, the NPOL lines clearly demarcate the difference in polarity between the relatively small ligands, the most polar of them being methotrexate (purple line NPOL = 15), and the larger more polar "biologics" (Oxytocin, Leuprolide) that occupy the lines with NPOL > 21, with a gap where there are no compounds. Second, all along the maps (both Figs. 5.4A and B) distinct lines are evident that correspond to the same small ligand (e.g., Methotrexate) whose affinity has been measured under different conditions or for different targets (human, *Escherichia coli*, mouse) depending on the cell assays or

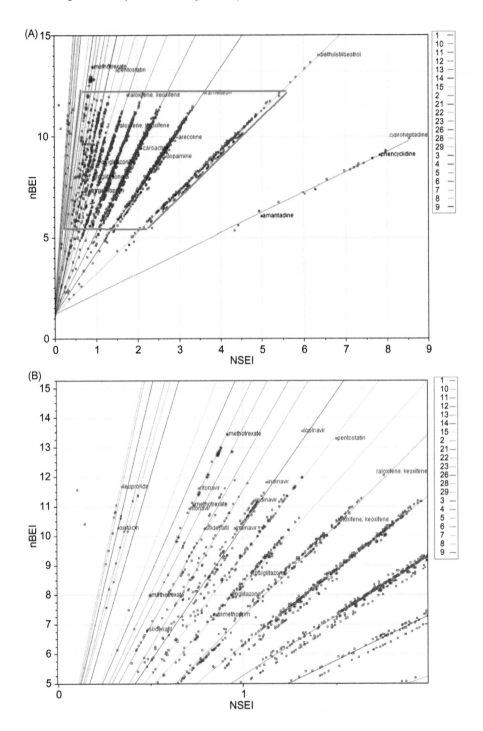

animal models used to develop the compounds. This simple example highlights the tremendous advantage of having a representation that is "scale invariant" in the sense that all the planes have the same direction lines, given by the NPOL parameter at any scale. There will be more illustrations and discussion of this type of efficiency planes in relation to different targets in Part III.

5.3 MAPPING OF BIOACTIVE PEPTIDES

The difficulties of representing in the AtlasCBS framework molecules that are very polar have been hinted at earlier. The problems are compounded because at high polarity the slope lines are very close together and also because the different polarities of the compounds cannot be very well described simply by the NPOL (N + O) count. These problems are particularly severe for the mapping of bioactive peptides, which represent a very important part of CBS. Nonetheless, it is possible to represent this important class of bioactive chemical entities in the AtlasCBS scheme to pursue further analyses.

The 2009 version of ChEMBL contained ~23,000 peptides (or peptide-like) entities that were defined as having at least a single peptide (CO−NH) bond. We will present the mapping of these molecules in various planes, showing their distribution in the planes, and illustrate their distribution in relation to the targets to which they are related. A more detailed study would be necessary to draw any quantitative conclusions.

◀ *Figure 5.4 (A) Mapping of marketed drugs included in WOMBAT (2007). Mapping in the NSEI−nBEI plane of the AtlasCBS the content of all the drugs in the market from WOMBAT (2007). The database includes 7,722 affinity measurements (both IC$_{50}$ and K$_i$). This representation corresponds to 3,585 K$_i$ measurements to provide a more uniform sample. Certain drugs have been highlighted with the corresponding name and some targets have been noted with a different shape marker. Examples, blue triangles: muscarinic receptor; green squares: PPAR-γ; purple stars: dihydrofolate reductase, DHFR. Well-defined lines of symbols (e.g., along the lines of dopamine or sidenafil) correspond to affinity measurements for the same molecule in different assays or for different variants of the same target (same slope and intercept). The position of the compound along the line depends on the K$_i$ value for that target. The trapezoidal region enclosed within green lines demarcates the region of CBS where most (~90%) of the drugs map in this efficiency plane. Image prepared with the JMP9 (SAS Institute) package, which was used for the statistical fitting corresponding to the NPOL lines labeled on the right panel. (B) Mapping of marketed drugs included in WOMBAT (2007) (detail). (B) Close-up of the map shown in (A) enclosed within NSEI (0,2) and nBEI (5−15), focusing on the region where most entries lie. The gap between the smaller, less polar (up to methotrexate), molecules and the larger peptide-like therapeutics is apparent. NPOL lines were modeled as indicated on the right panel. Oxytocin is a cyclic peptide containing nine amino acids and has 24 polar atoms (N + O). Leuprolide is a nonapeptide with a MW ~ 1,200 and NPOL = 28. Changing the scale of the axes facilitates the analysis and mapping of very different parts of CBS with consistent directions (NPOL: slope, affinity: radial distance). Analysis performed and illustrated with JMP9 (SAS Institute).*

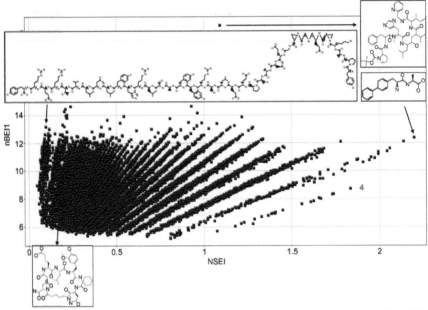

Figure 5.5 Overall mapping of the bioactive peptides contained within ChEMBL. The image maps the ~23,000 peptides included in ChEMBL (2009). It covers a range of 4–138 for NPOL although only 4–101 is represented. Some illustrative chemical structures have been shown. Courtesy of Dr. P.A. Bento (ChEMBL).

An overall view of the mapping of the peptide-like content of ChEMBL is presented in Fig. 5.5. The fan-like characteristic appearance of the NSEI–nBEI plane is well preserved and clear in the low NPOL values, but the lines are not so well resolved at high NPOL values. The lines are rather thick because there is a wide range of molecules with the same NPOL but expanded MW that extends beyond the central line due to the $\log_{10}(NHA)$ term. Of note is the very limited range of NSEI values in the abscissa that nonetheless covers all the included peptides. A change of scale and range in the plot could focus on certain (higher polarity) ranges of the plane for specific study. More importantly, it is possible to select a subset of the entire sample, i.e., NPOL = 7, and plot those compounds onto the SEI–nBEI plane. This plane is uniquely suited to separate molecules with subtly different polarities. As has been mentioned in Table 2.1 and shown in Appendix A, in the nBEI vs. SEI plane, the slopes of the lines are separated by one-hundredth of the PSA of the ligands (Fig. 5.6A and B).

Finally, it is possible to illustrate the distribution of bioactive peptides in CBS and relate them to the corresponding targets as has been

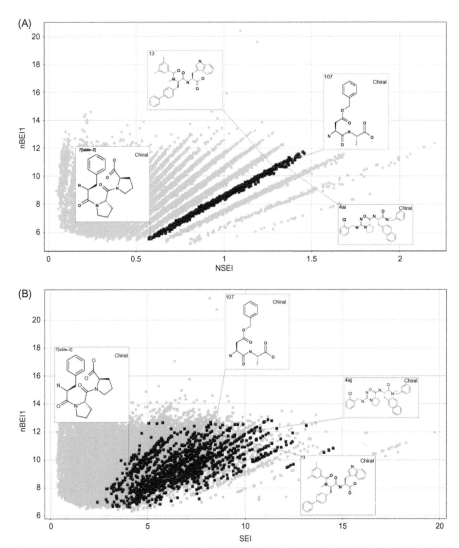

Figure 5.6 Selecting a particular NPOL line and displaying it with different variables. (A) The chemical entities having NPOL = 7 in panel (a) have been selected and displayed in the nBEI vs. SEI plane in panel (b) where the slope of the lines is PSA/100. (B) The fine separation of the physicochemical characteristics (i.e., polarity) in the angular coordinate can be appreciated. Some chemical structures are shown for reference. Figures prepared with Spotfire™, courtesy of Dr. P.A. Bento (ChEMBL).

shown in Fig. 5.7. Different colors have been used to identify the different "clusters" of peptides active against certain classes of biological receptors. As can be seen, there is an overlap among the different clusters associated to the various targets and these clusters are far from being as well demarcated as we might have anticipated (Fig. 3.1).

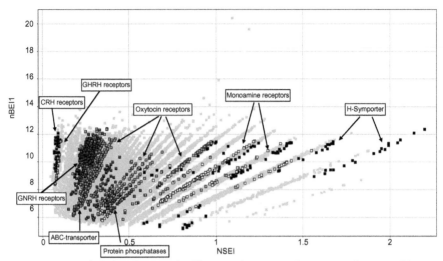

Figure 5.7 Mapping the biological space explored by various bioactive peptides. Mapping of a variety of bioactive compounds in relation to the class of biological molecules modulated by them. Note the very narrow range of NSEI (0 to ~2.25), the large variety of targets, the overlap and the broad extension in chemical space of each group of bioactive molecules. Colors as indicated. Prepared using Spotfire™. Courtesy of Dr. P.A. Bento (ChEMBL).

More detailed analyses will be required to extract significant insights from this representation since many of these affinity measurements include K_i and IC_{50}, and the different chemical entities could be structurally and functionally related.

5.4 DISCUSSION

Although they have to be regarded as early applications, the examples above highlight the advantages of having a unified representation of the content of SAR databases in the AtlasCBS framework. The highlighted efficiency plane, NSEI−nBEI, is easily interpreted as an initial mapping of the content of the databases in terms of polarity and affinity ranges, and can help to provide a quick visual representation of the compounds available, their distribution in polarity and affinity space toward their corresponding targets. A more quantitative analysis in the future could provide statistical (or probabilistic) estimates for drug-likeness as an additional coordinate (i.e., "altitude") that could map CBS in terms of high−low probability of success for further clinical development. Different planes, with complementary variables, or additional ones orthogonal to the ones described here could provide supplementary insights to guide drug discovery effectively.

Despite the high polarity and large number of atoms, bioactive peptides (i.e., biologics) can also be adequately represented in the AtlasCBS framework. Depending on the range of polar atoms and size, very polar areas of CBS can be studied in detail by a suitable choice of variables and ranges. Of particular interest is the plane defined by variables SEI−mBEI because the slope of the resulting lines is discriminated by one-hundredth of the PSA of the molecule.

A note of caution must be sound for the future. Natural products represent a unique subset of CBS that has not been considered here explicitly.[17] Because of their chemical complexity and polarity, these chemical entities will occupy the regions of the AtlasCBS where the slopes of the lines are very close together at high values of NPOL (Figs. 5.1, 5.4A and B). Attempts to combine chemical space with the richness of natural products and the corresponding biological targets are being made[18] and will be considered for future developments within the AtlasCBS framework. This issue will be studied in more detail in the future, but the flexibility of the presented framework allows the selection of very small areas of any efficiency plane by magnifying the scales, changing the variables, or selecting unique NPOL lines, as has been illustrated (Figs. 5.4A and B and 5.5−5.7).

Fragment-Based Strategies

The progressive abandonment of the natural products departments in large pharmaceutical companies was due, in part, to the lure of the high-throughput screening (HTS) approach combined with the advances in combinatorial chemistry. Rather than the painstaking effort of characterizing and identifying novel biological entities from bacterial, fungal, or even sea-dwelling organisms, the promise of finding active molecules against novel targets screening the large compound collections of the established pharmaceutical companies lured drug seekers into a different mindset. The concept sounded promising: develop a high-throughput assay and test the private chemical collections containing several hundred thousands of compounds against the novel target. The suggestions were that some compounds would be active and the corresponding chemical entities could be optimized readily for the novel targets. Limitations to this scheme appeared soon: (i) less than optimal pharmacokinetic properties; (ii) limited synthetic options and, more importantly, (iii) less than optimal binding of the preexisting compound to the novel target (limited binding). The concept of using smaller fragments for screening seemed to be more reasonable, followed by the linking of the fragments. It was like starting from scratch, optimizing each step along the way. This fragment-based (FB) drug-strategy concept was soon illustrated with a few dramatic examples using nuclear magnetic resonance (NMR) as the structural support: structure–activity relationship by NMR[19,20].

This concept is the latest incarnation of the classical structure-based drug design paradigm using crystallography or NMR as the supporting technology. The successes and perspectives of this methodology have been reviewed by some of the pioneers at Abbott[21]. The pressures to define the best (or more efficient fragments) highlighted the importance of the early concept of ligand efficiency as proposed by Hopkins and coworkers[22]. These early ideas and their most recent developments have been fully discussed in Chapter 2. How can the concepts related to ligand efficiency indices (LEIs) and the introduced mapping of AtlasCBS aid the FB approaches? Given the limitations of space, we will illustrate only two examples: (i) analysis of the content of fragment libraries and (ii) deconvolution of fragments and natural products. The AtlasCBS mapping has also been used in the novel field of poly-pharmacology but it will not be presented here. Details can be found in previous publications[3,16].

6.1 ANALYSIS OF THE CONTENT OF FRAGMENT LIBRARIES

The combination of two complementary indices (binding efficiency index (BEI), surface efficiency index (SEI); nBEI, NSEI; or NBEI, NSEI, or complementary pairs) in a Cartesian plane permits a 2D evaluation of the library content.

The concept has been already presented in the earlier chapters (Part I) but it is worth mentioning again. The combination in a Cartesian plane of two variables that share a common parameter in the definition allows visual insights that can help reinterpret the content of the elements in the map. In the case of the BEI–SEI pair (y,x), the ratio of the two variables, and thus the slope of the lines in the plot, depends only on the physico-chemical properties of the chemical entity (BEI/SEI = 10(PSA/MW)). As shown in previous publications and summarized earlier, the choice of the pair NSEI–nBEI (x,y), results in the target–ligand pairs being mapped along lines of slope of number of polar atoms, NPOL(= N + O). The important point in both cases is that the chemical entities *per se*, irrespec-tive of their affinity toward the specific ligands (K_i or equivalent), map along lines whose slopes are given by 10(PSA/MW) or NPOL, respectively.

These insights permit the "simulation" of results from screening experiments (either virtual or experimental) by assigning random affin-ities (for instance, as K_i values) to the library compounds. Calculation of the corresponding LEIs for compounds included in the fragment libraries under consideration permits a mapping of the chemical con-tent on the corresponding efficiency planes based on simulated affinity values. An example is presented in Fig. 6.1.

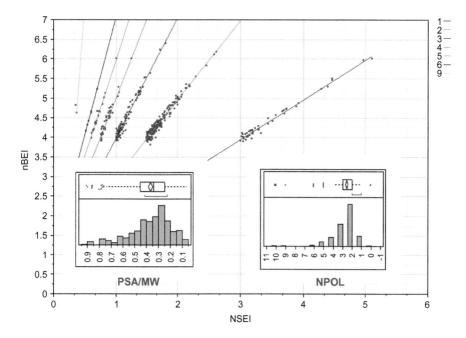

Figure. 6.1 Analysis of the content of chemical libraries in the NSEI–nBEI plane. Compounds included in the ActiveSight[24] library were assigned random K_i values (mM–μM range). The corresponding efficiency indices nBEI and NSEI were calculated and the content of the library displayed. This 2D representation, including the simulated K_i values, permits an exploration of the content of the library in chemical space (angular coordinate) and explores the regions of CBS accessible by the corresponding affinity values. Insets show a more conventional depiction of the content of chemical libraries using the distribution of the number of compounds with a different PSA/MW or NPOL (increasing right to left). Adapted from Abad-Zapatero and Blasi[16] and reproduced with permission. Copyright Molecular Informatics, 2011.

Figs. 6.1 and 6.2A compare the content of two libraries: a small library containing 384 compounds (ActiveSight)[25] and a more extensive Prestwick library containing 1,200 compounds that include 100% FDA-approved drugs[24]. Because the K_i values are assigned randomly, the efficiencies will map randomly along the different lines. The different chemical content of the libraries can be "read" in the distribution of values in the angular coordinate. The ActiveSight library (Fig. 6.1) contains a limited number of compounds with NPOL = 1–6, 9. There are only two compounds with NPOL = 9 and one with NPOL = 10 (excluded from the analysis). Looking at the "thickness" of the lines, one can see that only the compounds with NPOL = 2, 3 have any significant chemistry around the scaffolds. In contrast, Fig. 6.2A shows the content of a more extensive and "richer" library (Prestwick[24]) containing a vast majority of FDA-approved compounds. There is much more abundance of compounds with larger NPOL = 1–28, 33, 37, 40,

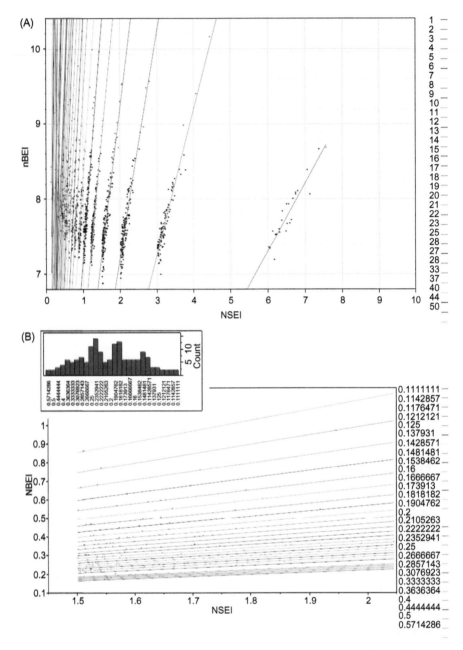

Figure. 6.2 Analysis of the content of the Prestwick library in two efficiency planes. (A) Mapping of the content of the Prestwick library (1,200 compounds)[24] in the NSEI–nBEI plane with random K_i values (μM–nM range). The wider range of NPOL values, including highly polar (NPOL > 6) molecules, is apparent. (B) Selected compounds with NPOL = 4 from the previous figure, represented in the NSEI–NBEI plane with slopes (NPOL/NHA) ranging from ∼0.1111 (4/36) to ∼0.5714 (4/7). Inset shows the distribution of NPOL/NHA ratios. Adapted from Abad-Zapatero and Blasi[16] and reproduced with permission. Copyright Molecular Informatics, 2011.

44, and 50. Although the more polar compounds seem to coalesce in the most vertical lines (higher slopes), a selection of the proper region of the plane and a change in scale (and variables) permits a detailed analysis.

For instance, Fig. 6.2B shows the "close-up" of the compounds corresponding to NPOL=4 in Fig. 6.2A and "spreads" them by using the (NSEI, NBEI) plane where the slope of the lines is given by NPOL/NHA and the polarity of these compounds is "magnified" and increases from NPOL/NHA = 0.111111 (4/36) to 0.57142857 (4/7) counterclockwise in the angular coordinate.

6.2 DECONVOLUTION OF FRAGMENT EFFICIENCIES

Two brief examples are presented to illustrate the value of a FB efficiency analysis in 2D planes rather than monitoring only the size-based optimization.

In a thought-provoking article, Hadjuk[1] reviewed the optimization path of 18 drug leads from 15 discovery programs at Abbott Laboratories in an attempt to extract optimization trends in the fragments making up the compounds. The data containing the affinity, chemical structures, and physicochemical parameters of the ligands were provided in the supplementary material of the publication, including the BEI values. The analysis using both SEI and BEI variables for the Bcl-xL target is summarized in Fig. 6.3. It is apparent that although the BEI remains approximately constant from the initial compound (6), the SEI is optimized from 5.5 to 7.3. Similar analysis of the other targets could provide further insights beyond the inferences made for the changes in BEI along.

Natural products are typically very complex chemical entities, reflecting the selection pressures affecting the structure−function pair through eons of time. The active compounds are typically a patchwork of fragments strung together and it is often difficult to identify the most efficient "head" for the purposes of further chemical synthesis and optimization by the medicinal chemist. Such a process was attempted with the natural product Argifin, a cyclopentapeptide that inhibits a wide variety of chitinases with activities ranging from micro- to nanomolar. The authors used crystallographic methods to identify the mode of binding of each fragment that they considered and

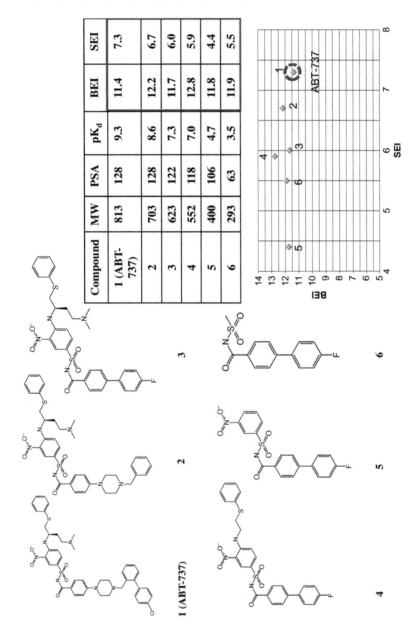

Figure. 6.3 Deconstruction of a FB ideal optimization. Summary of the ideal deconvolution of a Bcl-XL inhibitor in fragments. The values of pK$_d$, MW, PSA, and BEI were available from the supplementary information. The additional SEI variable was calculated and the values plotted in the lower right panel. The structures of the sequence are presented on the left with the corresponding modifications highlighted in blue. ABT-737 indicates compound selected for further development. Adapted from Hajduk[1].

Compound	MW	PSA	pK$_d$	BEI	SEI
1 (ABT-737)	813	128	9.3	11.4	7.3
2	703	128	8.6	12.2	6.7
3	623	122	7.3	11.7	6.0
4	552	118	7.0	12.8	5.9
5	400	106	4.7	11.8	4.4
6	293	63	3.5	11.9	5.5

measured affinity values for each one of them ranging from the smallest dimethylguanylurea to the complete Argifin molecule. Their analysis demonstrated that the smallest fragment, corresponding to approximately one quarter of the molecular mass and showing the weakest affinity ($IC_{50} \sim 500$ μM), had the highest efficiency of binding on a per mass basis[25]. Adding the polarity component to this analysis (Fig. 6.4), it can be shown that the smallest fragment also has the most efficient binding per unit of polar surface area (PSA) and thus optimizes the two efficiency indices. Indeed, this minimum fragment would be an excellent starting unit for further synthesis of chitinase inhibitors that could lead to drug-like compounds with a wide range of chemotherapeutic potential.

6.3 DISCUSSION

FB approaches are becoming more widespread within the drug-discovery community. Its successful application requires a substantial commitment of structural, physicochemical, chemical, and personnel resources to the project. Initially, it was proposed as a kind of panacea to overcome the limitations of lead discovery based on the results of HTS campaigns. Naturally, after being applied to more targets during the last decade its power and limitations have been diagnosed[21].

The FB methodology presented two important issues that had to be addressed: (i) design and diversity of the fragment libraries to be used in the screening and (ii) define and quantify the variables that should be used to identify the most efficient fragments and the corresponding ones to monitor the optimization process. The examples presented here have illustrated how to use the concepts of mapping chemicobiological space (CBS) in efficiency planes to address these two problems.

As has been illustrated, AtlasCBS representation can also be used to represent the content of chemical libraries (fragments or of any composition) by using randomly assigned affinity measurements. The framework permits these analyses by focusing on the angular coordinate of the maps, that it is solely dependent on the chemical composition. Brief examples have been presented on how to use these insights to compare the content of libraries in polarity space and also in exploring the affinity (radial) component accessible along the different lines of polarity. More elaborate analyses could be done by using the

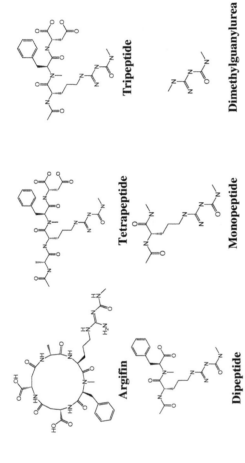

Argifin

Tetrapeptide

Tripeptide

Dipeptide

Monopeptide

Dimethylguanylurea

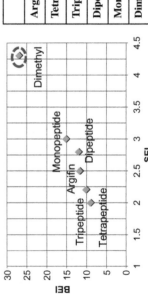

Compound	BEI	SEI	IC$_{50}$
Argifin	11.7	2.5	0.027
Tetrapeptide	9.1	2.0	4.3 ± 0.2
Tripeptide	10.2	2.2	5.1 ± 0.2
Dipeptide	12	2.8	12 ± 1
Monopeptide	15.3	3.0	81 ± 4
Dimethylguanylurea	27.2	4.3	500 ± 20

Figure 6.4 Efficiency dissection of the natural cyclic pentapeptide Argifin. Extension of the data provided by Andersen et al.[25] by adding the polarity-related efficiency (SEI). The least potent fragment (dimethylguanylurea) of the natural product Argifin turns out to be the most efficient fragment that optimizes both BEI and SEI (lower left panel).

different affinity planes, in particular SEI−BEI, where the slope of the lines is proportional to PSA/MW.

Typically, the analysis and comparison of chemical or fragment libraries have been done by examination and plotting of the distribution of the most important physicochemical properties of the molecules included in the library. Using those criteria, fragment libraries have been designed and are provided by several public and private chemical institutions or companies. These criteria are limited in that they do not consider the effect that the affinity toward the target would have in defining the region of CBS that will be explored (or accessible) in the screening experiments. The framework of the AtlasCBS concept presented here permits a true simulation of the screening process by assigning random affinity values to the chemical content of the library. Indeed, *a priori* it would be impossible to know which particular molecule would have any given affinity (efficiency) but the simulation experiment does permit to assess how far any chemical scaffold could go toward reaching drug-like efficiency.

The FB methodology gave a tremendous impetus to the notions of ligand efficiency in relation to identifying the most efficient ligands and predicting the expected gains in efficiency along the different stages of lead identification and optimization. Most of the analyses have been based on monitoring the efficiency gains in the size-related efficiency of the fragments and the resulting leads from the linking of the best fragments. In the study illustrated here, we have used the combination of two efficiency indices (typically BEI and SEI) to show that the fragment optimization process is better described as the optimization of these two types of efficiencies. This 2D optimization process is more suited as a strategic plan than the size-related one alone.

CHAPTER 7

Navigating in Chemicobiological Space

7.1 TRAJECTORIES IN CHEMICOBIOLOGICAL SPACE

The notion of "trajectory" to follow the optimization of fragment-based efforts was introduced by Willem and Nissink[26] as a result of plotting ligand efficiency (LE) (or SILE) vs. nonhydrogen atoms (NHA). A similar notion was suggested by Hadjuk[1] in his deconstruction of several lead compounds. However, both suggestions emphasized only the optimization of the size-efficiency variable. An attractive outcome of the unified definitions of ligand efficiency indices (LEIs) and their use in pairs as descriptive variables is that the process of "drug discovery" can be represented in 2-D maps in chemicobiological space (CBS). Motion in space immediately suggests the concept of trajectories as one moves from one point in CBS to the next one. Currently, we emphasize planar trajectories but n-dimensional trajectories might be used in the future. The location of compounds and series in the efficiency plane SEI−BEI (x,y) for the drug design efforts for human PTP1B (hPTP1B) was first presented in 2007[2]. Even though the chemical compounds for the active site of hPTP1B were very polar (number of polar atoms (NPOLs) >7), it was still possible to discern by their position in the plane that some compounds were superior to others by their larger values of the SEI variable (related to polar surface area (PSA) of the compounds).

The concept of trajectory of a drug-discovery project in CBS was introduced in a later publication, using also compounds from hPTP1B and representing them in the more intuitive and direct NSEI−nBEI plane[3] (Suppl. material, figs. S2a−d). However, given the highly polar nature of the active site pocket in hPTP1B, the compounds synthesized in different laboratories with rather different scaffolds never reached acceptable bioavailability and potency, which is reflected in the low values of the polarity-related efficiencies (low values of nBEI, NSE). In contrast, the conformationally-locked sulfonamide series for carbonic anhydrase II (CAII) achieved a marketable drug (Brizolamide, Azopt^{TM}) for the treatment of glaucoma by adding a couple of polar atoms to the scaffold while approximately retaining the efficiency per size (Fig. 7.1A and B). This trajectory is hypothetical as it is based on the deposited data of the Protein Data Bank (PDB) and the time of publication (deposition) might not reflect the date of conception, design, synthesis, and testing[27]. However, it illustrates the concept of trajectory using public domain data for a successful drug-discovery program. The reader is encouraged to explore independently the available data currently in BindingDB using the AtlasCBS application using the PDB entry codes shown in Fig. 7.1A (see Part III for examples).

The series presented in Fig. 7.1A and B illustrates this concept by highlighting the characteristics of the "motion" in the efficiency plane depending upon the substitutions in the ligand and resulting affinities. Compounds with the same number of NPOL atoms will move along the same line in the efficiency plane, upward if they gain in potency and downward if they are less potent. Depending upon the intersect term ($\log_{10}(NHA)$) the position of the compounds will depart away from the line but as it is a logarithmic term, it is much smaller and will depart from the line only slightly. Compounds designed with a similar scaffold with one additional polar atom (N,O) will "jump" to the line in the immediate left (counterclockwise), jumping one line for each additional polar atom. This is independent of where the change in the chemical scaffold has taken place. If polar atoms are removed from the scaffold, the compound location in the map will be moved to the corresponding lines in a clockwise manner. The exact location of the compounds along the corresponding lines will depend on the corresponding assay and affinity values (K_i, IC_{50}, K_d). If the affinity values are obtained with a robust assay for all the compounds, the positions in the efficiency plane are reliable and can be used to compare the compounds on the basis of LEIs in both dimensions. Estimates of K_i values by various theoretical docking protocols could also be obtained and the corresponding positions could also provide a "reasonable" comparison that could be used as guide.

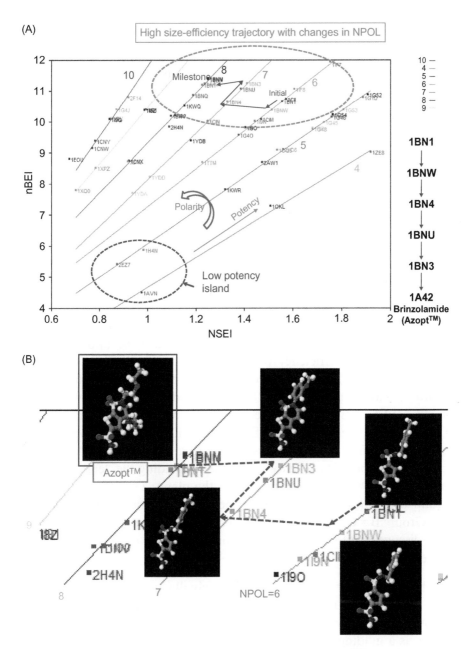

Figure 7.1 Hypothetical trajectory of compounds for CAII leading to Brinzolamide (Azopt^TM). (A) The PDB entries containing CAII complexes with relevant bound ligands were extracted from PDBBind. The different efficiency indices were calculated and the results plotted and analyzed with JMP9 (SAS institute). The compounds clustered in lines of NPOL = 5−10 (see panel). A hypothetical trajectory (right panel) leading to the marketed drug Azopt^TM (Brinzolamide) has been suggested on the basis of the published work[27]. (B) Close-up of the efficiency plane NSEI−nBEI presented in (A), in the proximity of different, highly efficient, CAII inhibitors. The directional NPOL lines are drawn for NPOL = 6−9. A hypothetical trajectory leading to the marketed drug Azopt^TM (Brinzolamide, an antiglaucomic agent) has been drawn and annotated with the chemical structures of the corresponding inhibitors. The PDB accession codes for the corresponding target−ligand complexes label each point[27].

This is probably the appropriate moment to relate the trajectories to the concepts of "activity cliffs" and "activity ridges" presented earlier in relation to mapping activities and chemical changes[28,29]. The appearance of the NSEI−nBEI maps make it very clear that dramatic shifts can take place upon adding or removing one or several polar atoms (NPOL = N + O) to a scaffold, particularly when the total number of them is rather small. The slope of the lines that contain these compounds will change by relative large amounts (i.e., ± 1, ± 2) and even though the exact location would depend on the affinity, the distances in the efficiency plane could be substantial. Changing the number of polar atoms of the hit or lead can have dramatic effects not only in potency but also in physicochemical properties. For drug discovery, a more effective way of looking at activity discontinuities ("activity cliffs" or ridges, in the current parlance) would be to consider rigorous distances in other variables, not only affinities. The efficiency planes presented here could be explored in the future.

This concept of trajectory in CBS is also germane to the efficiency plane (SEI−BEI) or any other plane and can also be applied to the concepts of group efficiency (GE) quite intuitively[30]. The preexisting scaffold (Group A) will have a certain $(PSA/MW)_A$ and therefore given its affinity can be placed along the corresponding line $(10(PSA/MW))_A$ on the (SEI, BEI) plane. Incorporation of the additional group (Group B with characteristics PSA/MW_B) will change the corresponding properties of the resulting compound to a value of $10(PSA/MW)_{AB}$ and the corresponding affinity value, thus providing the "end point" of the Group A > Group AB motion in the plane. The effect of adding Group B to Group A can be described as a "vector addition" Group A + Group B = Group AB in the plane(s) of choice used to describe drug-discovery effort. Similar arguments can be made of a similar process being represented in the NSEI−NBEI plane or others.

This notion of trajectory was extensively illustrated by extending the data published by researchers at Merck-Serono in their series against the well-known target 11β-HSD1[4,5]. The project originally used BEI values as a guiding criterion to optimize their series and they provided the corresponding values in the SAR tables of the original publications. After the publication of the extended definitions of LEIs, the expanded framework (efficiency per size and polarity) provided a "natural" way to represent the drug-discovery project, with their different series in various efficiency planes. This resulted in the comparison of trajectories for three different series (azaindoles, pentanedioic acids, and spirocarboxamides) in relation to "reference compounds" (Fig. 7.2). For clarity, the four different sets have been represented in four panels with the same section of the NSEI−nBEI coordinates. Thus, it is possible to compare the relative merits of the three chemical series in relation to the reference compounds (Fig. 7.2). As can be easily seen, the spirocarboxamides series is superior to the other two in terms of affinity/size and affinity/polarity. This is clearly appreciated because this chemical series maps much further to the right in the plane and the best compound of the spirocarboxamide series is also more efficient than the best compound of the azaindole series (nBEI > 8.5).

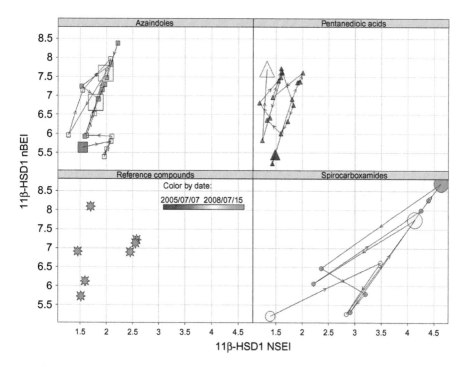

Figure 7.2 Overall trajectories of Merck-Serono compounds for 11β-HSD1. Representation on the NSEI−nBEI efficiency plane of three chemical series: azaindoles (upper left), pentanedioic acids (upper right) and spirocarboxamides (lower right), and reference compounds (lower left). The x,y axis ranges and scales are the same. Note how the spirocarboxamides series is far superior to the other three (including reference compounds) in the horizontal direction (polar efficiency). Size of the markers is proportional to potency and color relates to the date as indicated 2005/07/07 (red) to 2008/07/15 (green)[4,5,37]. The line connecting the starting and the end compounds would be the vectorial addition of the different steps and can be analyzed as illustrated in the following section. Reproduced with permission from Reference[37]. Copyright Molecular Informatics, 2011.

A more detailed view of the evolution of the different series in chemical, biological, and time sequence can be appreciated in Fig. 7.3A–C (azaindoles, pentanedioic acids, and spirocarboxamides, respectively). Figure 7.4 depicts the mapping of a similar spirocarboxamides series[4] in the efficiency plane and also the corresponding compound structures. The movement of the chemical series in the NSEI–nBEI plane it is easy to follow based on the NPOL number (N + O) of the structures. The selected compound for further development (13) is in the upper right hand of the diagram, where the two variables are optimized (Fig. 7.4).

In addition, because the dates of compounds synthesis (or testing in assays) were also stored in the internal database, it was also possible

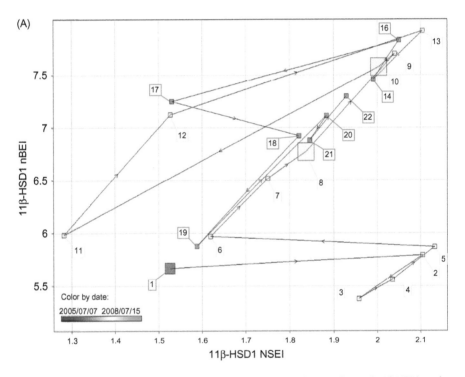

Figure 7.3 Time and efficiency trajectories of three series of compounds directed toward 11β-HSD1 in the NSEI–nBEI efficiency plane. (A) Azaindole series: the numbers correspond to the compound numbers discussed in the original publication. (B) Pentanedioic series: this represents an extensive series of compounds whose best leads in size and polarity (7, 18, 19, 21) were still far from the spirocarboxamide (Fig. 7.3C). Size of the triangles is proportional to affinity[37]. Color changes correspond to the timeline as indicated. (C) Spirocarboxamide series: mapping of the optimization trajectory for the spirocarboxamide series from the initial hit compound 1 to the optimized 11. Compounds are numbered as in the original publications[4,5]. Further details in relation to these three trajectories can be found in reference[37].

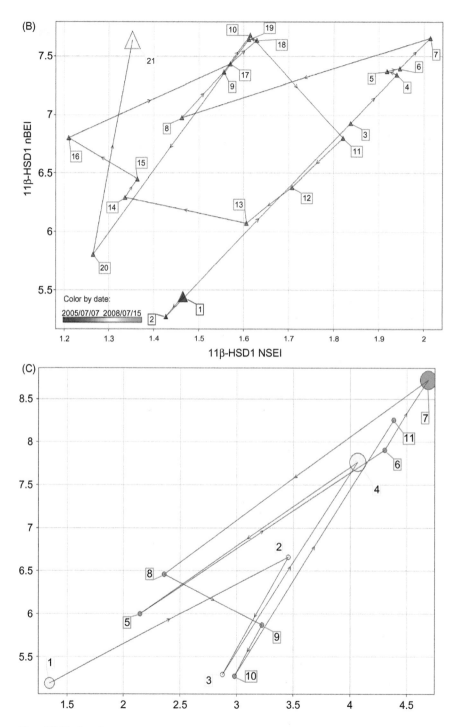

Figure 7.3 (Continued)

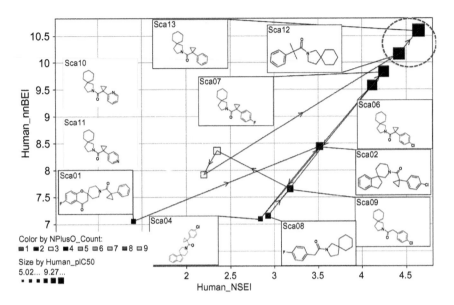

Figure 7.4 Efficiency trajectory in the NSEI–nBEI plane for the spirocarboxamide series. Trajectory of the spirocarboxamide series in the NSEI–nBEI efficiency plane. The jumps across the lines of different slopes (e.g., between compounds 1 and 2) are due to changing the number of polar atoms (4 vs. 2). Changes along the lines, for instance compounds 7, 9, and 12, are due to differences in affinity (within the same line NPOL = 2); compounds 12, 13 are the most efficient (red dashed oval). Sizes of the squares is proportional to affinity. This trajectory corresponds to the compound series described by Lepifre et al.[4,37]. Chemical structures are highlighted. Color of the symbols within the lines are color coded as indicated relating to the NPOL count. Figs. prepared with Spotfire[TM].

and indeed interesting to draw "trajectories in time" to examine parameters such as "duration of project," "time to optimization," and others. Time is depicted by the progression of color from red to green, as indicated (Fig. 7.3A–C).

It is suggested that the use of trajectories in efficiency planes (sections) within the vast biological space, as illustrated in Fig. 7.3, has only began and will, most likely, prove to be a very useful way of portraying the drug-discovery effort. In addition, by computational and statistical analyses it might be possible in the future to compare rigorously various trajectories for different series or alternative strategies, in terms of complexity of chemical synthesis, feasibility, and time required for optimization. With additional experience and data, the derivation or suggestion of optimal trajectories might be within our reach in a not too distant future. Most likely, success in drug discovery will not be a "numbers game" any more, but rather an efficient endeavor with a higher probability of going from beginning to end in a predictable manner.

7.2 THE SEARCH FOR OPTIMAL TRAJECTORIES

A pressing question that is critical to the application of these concepts quantitatively is whether they would have robust predictive value to optimize the drug-discovery process based on an objective, numerical, or at least statistical basis. Much work remains to be done to achieve this goal. Nonetheless, to draw some conclusions for the future application of these concepts, we have chosen to reexamine a recent work that used a dataset of 60 lead−drug pairs in the vector framework based on LEIs that provide some initial insight into this problem[31]. The set included results from 40 different companies, including enzyme targets (23) and receptors (16) with two drugs approved within 1978−1990 and the rest more recently (1991−2008). The goal was to illustrate the method and also to see if any meaningful visual trends can be extracted that could be given a more robust statistical formulation in the future. This preliminary analysis suggests that besides the graphical appeal of the proposed approach, the numerical framework permits to extract suggestive trends that may be confirmed by larger lead−pair datasets so as to set the basis for a more robust (numerical or statistical) optimization of the lead-to-drug process. This analysis was done while I was at the Platform of Drug Discovery (Parc Cientific Barcelona, Spain) with the valuable assistance of Daniel Blasi.

The 60 (lead/drug) pairs analyzed by Perola[31] were put into the LEIs framework. Additional data were kindly provided by Dr. Perola who had added the BEI, SEI values for all the pairs. The SMILES strings for all the pairs were added and from them the corresponding NPOL (N + O) and NHA were added. Using these values, the additional LEI pairs, namely nBEI, NSEI, and NBEI, NSEI were calculated using the definitions (Table 2.1). Different efficiency planes were examined to extract optimization trends from the combined variables simultaneously. The ligand lipophilicity index (LLE) index was also incorporated into the vectorial framework (Eq. (2.10), Table 2.1).

The most important results are summarized in Figs. 7.5 and 7.6. Figure 7.5 depicts a vectorial representation of the "migration" of the lead-to-drug vectors in the efficiency plane defined by NSEI−nBEI (x,y). Each compound (lead and drug) was mapped in the efficiency plane, and thus a vector can be drawn with the tail on the "lead" (blue circles) and the arrowhead in the "drug" (red squares). This representation clearly shows how the optimization process results in a "vector" in

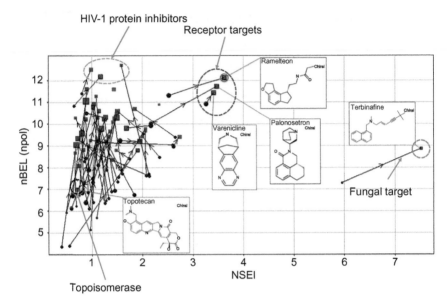

Figure 7.5 Vectors (lead > drug) in the NSEI−nBEI efficiency plane for the 60 lead−drug pairs included in the Perola dataset. The 60 lead−drug pairs analyzed by Perola[31] in the graphical representation including size (nBEI) and polarity (NSEI). Blue circles correspond to the lead compound (NSEI, nBEI)$_L$ and the red squares to the drug (NSEI, nBEI)$_D$. The size of the symbols is proportional to the LLE (−4.5−12.76). The arrowheads mark the direction of the vector lead > drug. Most of the drugs are found within a region bounded by 0 < NSEI < 2, 8 < nBEI < 11. A clear visual trend of an "upward" displacement in the size direction can be appreciated. Less prominent but also appreciable are shifts in the horizontal direction. A more quantitative analysis is presented in Fig. 7.6. A few drugs are shown for reference. Several "islands" where a few drugs related to specific targets seem to cluster have been highlighted with dashed ovals. Blue: HIV-1 protease; Red: several receptor-related drugs, for example, Varenicline: nicotinic receptor α4β2; Green: Terbinafine, very lipophilic antifungal agent inhibiting ergosterol biosynthesis; Black: Topotecan occupies a unique position in the map since it is a DNA-binding compound (topoisomerase target). Portions of this work were presented at the 18th European QSAR symposium (Rhodes, 2010) as a poster presentation (C. Abad-Zapatero, D. Blasi, J. Wass).

efficiency space. Similar plots were produced in other efficiency planes (SEI−BEI) and (SEI−LLE), but the visual impression was not as striking, even though statistical analysis might in the future reveal other numerical trends. Several conclusions can be drawn from this visual representation.

First, overall the optimization process results in vectors that move approximately "upward" and predominantly to the "right" of the plane. Second, the value of the LLE index, represented by the size of the symbols (blue circles vs. red squares), seems to increase. Optimization of certain targets seems to move compounds toward certain limited areas of the efficiency space. Various regions have been highlighted for compounds related to different types of targets: HIV-1 protease (Lopinavir), melatonin receptors (Ramelteon), fungal

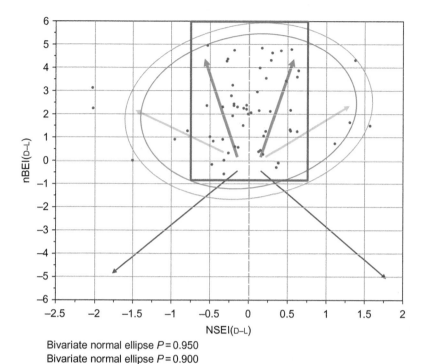

Bivariate normal ellipse $P = 0.950$
Bivariate normal ellipse $P = 0.900$

Figure 7.6 Quadrant distribution of the lead > drug vectors. The different vectors in size-efficiency (nBEI(D–L)) and polarity-efficiency (NSEI(D–L)) have been plotted for the 60 lead–drugs pairs analyzed by Perola[31] and represented in a quadrant depiction. Vectors above the horizontal line (different between nBEI(D)−nBEI(L) > 0) show the direction and amount of the change in going from lead to drug in the efficiency/size. Similarly, points to the right of the NSEI(D–l) = 0 denote vectors with a positive increase in going from leads to drugs in polarity space. Ellipsoids enclosing 95% (red) and 90% (green) of the dataset have been annotated for reference. The vectors correspond to the pairs and variables depicted in Fig. 7.5. Figures 7.5 and 7.6 prepared with Spotfire^{TM} and JMP9 (SAS Institute).

squalene epoxidase target (Terbinafine), and Topotecam^{TM} (a topo-isomerase inhibitor) that binds to DNA.

Figure 7.6 represents in a "quadrant" diagram the trends in the change of the modulus and direction of the vectors resulting from the "lead-to-drug" optimization process. As can be seen, practically no optimization of the vectors results in the decreasing of the efficiency per size (nBEI variable). The vast majority increases quite significantly (as much as 5 units in certain cases). For only four pairs the nBEI (D–L) vector component is negative and even in those cases barely so (about 0.5 units). Interestingly, the changes in polar composition (NSEI variable, NSEI (D–L) vectors) are rather small and limited, even though they go in both directions. With a few exceptions, most drug–lead pairs only differ by +/− 0.5 units in the NSEI component (46/60).

Although not fully quantified yet, we feel that the analysis of these trends presented in a 2-D frame (efficiency/size and per polarity) in a graphical representation permits an easier grasp of the optimization process and related to variables that are known to have strong weight in the optimization process. Further examples and more robust statistical analysis of larger samples should permit extracting solid numerical trends that could be used in the future to guide the drug-discovery process on a sounder basis. Also, the different efforts to provide size-independent[26,32] efficiencies or the work related to the separation of enthalpic and entropic terms of efficiency[33-35] would be important to provide reliable theoretical affinities. Finally, the most recent estimates of drug-likeness (QED, Table 2.2)[36] could provide "figures-of-merit" to drive the search for optimal trajectories in CBS and facilitate the robust prospective charting of "journeys" in drug discovery.

7.3 DISCUSSION

The notion of "trajectories" in hit-to-lead or lead-to-drug progressions is not new. This concept has been mentioned in previous publications, and it is becoming more accepted in the context of drug-discovery efforts[1,21,26]. Our emphasis has been in describing trajectories in "efficiency planes" or in "efficiency space" as opposed to monitoring only migration in chemical−property space or in relation to changes in binding affinities. We theorize that the critical issue in any kind of trajectory description is to find the best set of variables that describe the trajectory and, in this case, optimize the drug-discovery process with the most robust "figures-of-merit." After the predominant emphasis on affinity (potency) that has driven the process for many years, the jury is still out as to which (and how many) would be the most reliable guideposts to mark our progress and effectiveness.

A word of caution is also appropriate in the context of the metaphor of sea navigation as related to navigating CBS. The charts and accurate maps with precise latitude and longitude readings, providing a reliable depiction of the Earth physical features and making navigation a science more than "a perilous journey," did not happen overnight. Prior to the exquisitely detailed maps resulting from the large number of surveying expeditions all over the world by the British Navy, the seafarers of the Middle Ages relied on "portolan charts." They can be described as "local" maps derived from the detailed observations of ship pilots and captains using wind directions and crude compass settings as markers. These were early maps characterized by a crisscrossing of lines emanating from certain points with markings corresponding to "rhumb" lines. They had reasonably accurate descriptions of the shorelines and provided wind directions (and later compass settings) for relatively simple journeys, setting a course along a "rhumb" line or loxodrome: the course making a constant angle with any meridian of longitude. In a way, the NPOL lines that have appeared naturally in the NSEI–nBEI planes remind me of those "directional" lines. I surmise that we are still in these early times of charting and navigating through CBS. We have a long way to go to chart our course safely on the long journey of discovering drugs from hits, to leads and to drugs; and the safe peninsulas (or even islands) of successful drugs are still not very clear in our charts. Examples in Part III will add further insights into the use of these maps to aid drug discovery using practical examples.

REFERENCES FOR PART II

1. Hajduk PJ. Fragment-based drug design: how big is too big? Journal of medicinal chemistry. 2006; **49**(24): 6972−6.

2. Abad-Zapatero C. Ligand Efficiency Indices for effective drug discovery. Expert opinion in drug discovery. 2007; **2**(4): 469−88.

3. Abad-Zapatero C, Perisic O, Wass J, Bento AP, Overington J, Al-Lazikani B, et al. Ligand efficiency indices for an effective mapping of chemico-biological space: the concept of an atlas-like representation. Drug discovery today. 2010; **15**(19−20): 804−11.

4. Lepifre F, Christmann-Franck S, Roche D, Leriche C, Carniato D, Charon C, et al. Discovery and structure-guided drug design of inhibitors of 11β-hydroxysteroid-dehydrogenase type I based on a spiro-carboxamide scaffold. Bioorganic and medicinal chemistry letters. 2009; **19**(13): 3682−5.

5. Roche D, Carniato D, Leriche C, Lepifre F, Christmann-Franck S, Graedler U, et al. Discovery and structure−activity relationships of pentanedioic acid diamides as potent inhibitors of 11β-hydroxysteroid dehydrogenase type I. Bioorganic and medicinal chemistry letters. 2009; **19**(10): 2674−8.

6. Lipinski CA, Lombardo F, Dominy BW, Feeney PJ. Experimental and computational approaches to estimate solubiliy and permeability in drug discovery and development settings. Advanced drug delivery reviews. 1997; **23**: 3−25.

7. Ertl P, Rohde B, Selzer P. Fast calculation of molecular polar surface area as a sum of fragment-based contributions and its application to the prediction of drug transport properties. Journal of medicinal chemistry. 2000; **43**(20): 3714−7.

8. Cruciani G, Pastor M, Guba W. VolSurf: a new tool for the pharmacokinetic optimization of lead compounds. European journal of pharmaceutical sciences: official journal of the European federation for pharmaceutical sciences. 2000; **11**(Suppl 2): S29−39.

9. Veber DF, Johnson SR, Cheng HY, Smith BR, Ward KW, Kopple KD. Molecular properties that influence the oral bioavailability of drug candidates. Journal of medicinal chemistry. 2002; **45**(12): 2615−23.

10. Congreve M, Carr R, Murray C, Jhoti H. A "rule of three" for fragment-based lead discovery?. Drug discovery today. 2003; **8**: 876−7.

11. Kubinyi H. Drug research: myths, hype and reality. Nature reviews drug discovery. 2003; **2**(8): 665−8.

12. Wenlock MC, Austin RP, Barton P, Davis AM, Leeson PD. A comparison of physiochemical property profiles of development and marketed oral drugs. Journal of medicinal chemistry. 2003; **46**(7): 1250−6.

13. Abad-Zapatero C. A Sorcerer's apprentice and the Rule of Five: from rule of thumb to commandments and beyond. Drug discovery today. 2007; **12**(23/24): 995−7.

14. Zhang M-Q, Wilkinson B. Drug discovery beyond the "rule-of-five". Current opinion in biotechnology. 2007; **18**(6): 478−88.

15. Sepkowitz KA. One hundred years of Salvarsan. The New England journal of Medicine. 2011; **365**(4): 291−3.

16. Abad-Zapatero C, Blasi D. Ligand efficiency indices (LEIs): more than a simple efficiency yardstick. Molecular informatics. 2011; **30**(2−3): 122−32.

17. Lachance H, Wetzel S, Kumar K, Waldman H. Charting, navigating, and populating natural product chemical space for drug discovery. Journal of medicinal chemistry. 2012; **55**(13): 5989−6001.

18. Bon RS, Waldmann H. Bioactivity-guided navigation of chemical space. Accounts of chemical research. 2010; **43**(8): 1103–14.

19. Shuker SB, Hajduk PJ, Meadows RP, Fesik SW. Discovering high-affinity ligands for proteins: SAR by NMR. Science. 1996; **274**(5292): 1531–4.

20. Hajduk PJ, Sheppard G, Nettesheim DG, Olejniczak ET, Shuker SB, Meadows RP, et al. Discovery of potent nonpeptide inhibitors of Stromelysin using SAR by NMR. Journal of American chemical society. 1997; **119**: 5818–27.

21. Hajduk PJ, Greer J. A decade of fragment-based drug design: strategic advances and lessons learned. Nature reviews drug discovery. 2007; **6**(3): 211–9.

22. Hopkins AL, Groom CR, Alex A. Ligand efficiency: a useful metric for lead selection. Drug discovery today. 2004; **9**(May): 430–1.

23. ActiveSight. ActiveSight fragment library optimized for crystallography. 2007. Available from: <http://www.zenobiatherapeutics.com>.

24. Wermuth CG. Selective optimization of side activities: another way for drug discovery. Journal of medicinal chemistry. 2004; **47**(6): 1303–14.

25. Andersen OA, Nathubhai A, Dixon MJ, Eggleston IM, van Aalten DM. Structure-based dissection of the natural product cyclopentapeptide chitinase inhibitor argifin. Chemistry and biology. 2008; **15**(3): 295–301.

26. Willem J, Nissink JW. Simple size-independent measure of ligand efficiency. Journal of chemical information and modeling. 2009; **49**(6): 1617–22.

27. Boriack-Sjodin PA, Zeitlin S, Chen HH, Crenshaw L, Gross S, Dantanarayana A, et al. Structural analysis of inhibitor binding to human carbonic anhydrase II. Protein science. 1998; **7**(12): 2483–9.

28. Peltason L, Iyer P, Bajorath J. Rationalizing three-dimensional activity landscapes and the influence of molecular representations on landscape topology and the formation of activity cliffs. Journal of chemical information and modeling. 2010; **50**(6): 1021–33.

29. Stumpfe D, Bajorath J. Exploring activity cliffs in medicinal chemistry. Journal of medicinal chemistry. 2012; **55**(7): 2932–42.

30. Verdonk ML, Rees DC. Group efficiency: a guideline for hits-to-leads chemistry. ChemMedChem. 2008; **3**(8): 1179–80.

31. Perola E. An analysis of the binding efficiencies of drugs and their leads in successful drug discovery programs. Journal of medicinal chemistry. 2010; **53**(7): 2986–97.

32. Reynolds CH, Tounge BA, Bembenek SD. Ligand binding efficiency: trends, physical basis, and implications. Journal of medicinal chemistry. 2008; **51**(8): 2432–8.

33. Reynolds CH, Holloway KM. Thermodynamics of ligand binding and efficiency. ACS medicinal chemistry letters. 2011; **2**: 433–7.

34. Garcia-Sosa AT, Hetenyi C, Maran U. Drug efficiency indices for improvement of molecular docking scoring functions. Journal of computational chemistry. 2009; **31**(1): 174–84.

35. Ferenczy GG, Keseru GM. Enthalpic efficiency of ligand binding. Journal of chemical information and modeling. 2010; **50**(9): 1536–41.

36. Bickerton GR, Paolini GV, Besnard J, Muresan S, Hopkins AL. Quantifying the chemical beauty of drugs. Nature chemistry. 2012; **4**(2): 90–8.

37. Christmann-Franck S, Cravo D, Abad-Zapatero C. Time trajectories in efficiency maps as effective guides for drug discovery efforts. Molecular informatics. 2011; **30**(2–3): 137–44.

PART III

Exploring CBS: Practical Examples Using the AtlasCBS Server

Specific examples are presented of the application of the AtlasCBS web server to map, analyze, and interpret the data of retrospective and prospective studies in terms of optimization of the efficiency of the ligands. Using a simple dataset extracted for neuraminidase, the reader will learn to interpret basic patterns in the efficiency planes and the mechanics of using the AtlasCBS tool. Thereafter, the vast amounts of data available for targets such as HIV-1 protease and HIV-1 reverse transcriptase are analyzed within the AtlasCBS application to illustrate the evolution of the drug-discovery efforts in terms of displacements through the efficiency planes. These initial examples are followed by efficiency analysis for other targets in various therapeutic areas and using different drug-discovery strategies: protein kinases (cancer), acetylcholinesterase (neurotransmission), opioid receptors (analgesia, addiction), lactic dehydrogenase A (fragment-based drug discovery), and catechol-o-methyl- transferase (virtual screening). The final example highlights the prospective use of the concepts learned in a lead optimization project related to rare amyloid diseases, targeting transthyretin. The goal is to stimulate the drug-discovery community to use these concepts and methods in their discovery efforts. Only the collective endeavor of developing and testing new concepts and tools will make us more effective.

CHAPTER 8

Sightseeing Chemicobiological Space

INTRODUCTION: SNAPSHOTS, POSTCARDS, VISTAS, AND NAVIGATIONAL CHARTS OF CBS

The previous chapters have presented the vastness of chemicobiological space (CBS) and proposed a certain algebraic framework introducing new variables that represent the chemical entities and their associated biomolecules on 2D charts. The map collection therefore constitutes an atlas-like representation of CBS. This part will be very different from the previous two. This part aims to provide a series of snapshots, postcards, vistas, and possibly navigational charts of certain portions of CBS. The space left to this small opus is too limited to present the complexity of that universe. However, with the collection presented, I intend to whet the appetite of both the naïve student and the seasoned investigator to visit and explore CBS in a novel way. The same way I did when I was trying to select views for this collection. In what follows, there are single snapshots of certain areas of CBS, brief postcards exposing a specific landscape for a target, broader vistas and panoramas, and, in certain cases, a more detailed nautical chart of the waters sailed by others looking for drug treasures for various biological targets. Because I used data already published or available in existing databases, the views are predominantly retrospective; but do not be deterred: one example uses the AtlasCBS approach in a prospective manner. As you browse or study these materials, think of what other views you might have selected and how you can apply these ideas to expedite your journeys through CBS. Whether you are seriously exploring CBS prospecting for riches, treasure islands, or whether you are doing it just to satisfy your own curiosity, try to see if the ideas presented previously, or the inferences obtained from this later material can help you in your journey. If they can, the work will have achieved its intended purpose. If not, we (the communal we) will have to continue our search for better charts and superior navigational instruments. Your experience and the future will tell.

8.1 GETTING STARTED: NEURAMINIDASE

Influenza, the "flu", is a serious respiratory ailment that has a long history of affecting human health and welfare. Epidemics caused by this pathogen have been recorded for centuries and of particular resonance is the 1918–1919 outbreak that killed an estimated 20–40 million people worldwide. In addition, certain strains of the avian flu (e.g., H5N1) can infect people and under certain conditions cause high mortality. During the last two decades, genetic, biochemical, and structural biology work has identified the neuraminidase molecule as a target for chemical intervention in the efforts against the influenza virus.

There is a significant amount of data in public-domain structure–activity relationship (SAR) databases on the neuramidinase (strains A, B, and various subtypes) and we will illustrate with a simple example: (i) how to access the data; (ii) how to plot it; and (iii) how to interpret the results in terms of the physicochemical properties of the important molecules that were developed to treat this viral infection.

BindingDB has three separate entries for this target: neuraminidase, neuraminidase A and neuraminidase B. The first tab contains data for both A and B viruses as well as various different strains, for instance, /Puerto Rico/8/34(H1N1). In most cases, there are affinity data for both K_i (3) and IC_{50} (14) values, and by mapping the two as separate entities one can see that the two assays give consistent results. If one chooses the neuraminidase A tab (A/Puerto Rico/8/34(H1N1)) target with IC_{50} values (blue) the efficiency map illustrates the drug-discovery efforts in two series of compounds related to sialic acid: tetrahydropyridazines and the carbocyclic analogs with the C2-substituted and C3-Aza (3 h) analogs illustrated in Fig. 8.1, culminating in GS4071, a potent inhibitor[1]. A faster way to upload some of the data would be to enter the protein data bank (PDB) access code 1A4G in the "Add PDB source" window, within the **Viewer** tab. This will directly upload the data corresponding to neuraminidase B with the compound Zanamivir highlighted (Fig. 8.2A). The code corresponds to the crystal structure of the complex between neuraminidase and Zanamivir. The other data could be added as indicated (Fig. 8.1).

Selecting the neuraminidase B target presents a narrow distribution of compounds (red) where Zanamivir is clearly differentiated as being the most efficient in size and also in polarity (x-axis) although by a narrower margin in the latter variable. Zanamivir was the first useful inhibitor of the neuraminidase enzyme with therapeutic effect, discovered in 1989 using target-guided and structure-based strategies. This compound has a large number of polar atoms ($N + O = 11$) and, not surprisingly, has poor bioavailability. It is administered by inhalation and consequently has limited clinical use.

Further to the right (Fig. 8.2B), with only six polar atoms, is Oseltamivir, the first orally active neuraminidase inhibitor developed by C.U. Kim at Gilead Science a few years later[1] that was marketed by Roche as the well-known antiflu medicine Tamiflu[TM]. It should be pointed out that Oseltamivir is the prodrug of GS4071 that is activated by the cleavage of the ethyl ester exposing the COOH group that interacts directly with the positive parts of the neuraminidase active site, in particular Arg224 (PDB entry 2HT8). It has been shown that GS4071 is the major metabolite (32−56%) of [2-acetyl-^{14}C]Oseltamivir observed in metabolic studies in rats after oral dosing[2].

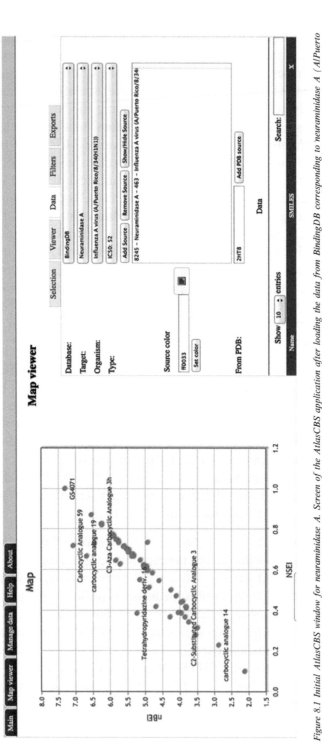

Figure 8.1 Initial AtlasCBS window for neuraminidase A. Screen of the AtlasCBS application after loading the data from BindingDB corresponding to neuraminidase A (A/Puerto Rico/8/34(H1N1)).

(A)

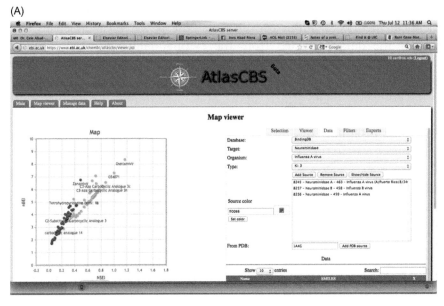

(B)

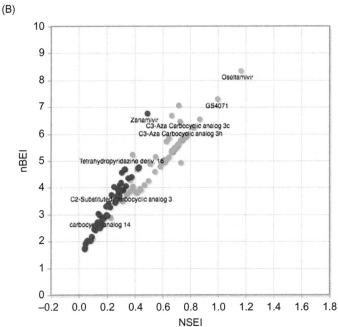

*Figure 8.2 (A) AtlasCBS window for neuraminidase compounds. Screen after the loading of three datasets from BindingDB containing data for neuraminidase. After selection, the compound structures and links to the original database can be seen using the **Selection** tab. (B) Close-up of Fig. 8.2A. Exported map image from the full screen in Fig. 8.2A using the **Exports** tab in the **Viewer** screen. The different compounds and the corresponding efficiencies along the path to the discovery of Zanamivir, GS4071, and Oseltamivir can be seen and the structures explored using the **Selection** window from the screens shown above.*

The efficiency analysis of these three important compounds in the chemical therapy of influenza illustrates the power of a direct inspection of the data extracted from the exiting database with a simple example. Leaving aside the issue of the activation after cleavage, the mapping of these two active compounds (Zanamivir and Oseltamivir) in the efficiency plane clearly shows how Oseltamivir, the orally active compound, is more efficient in the two variables considered nBEI and NSEI. In addition, the displacement toward the right of the active compound (GS4071) and the orally administered compound prodrug (Oseltamivir) is also consistent with these compounds being more drug-like, with a lower number of polar atoms. Note that Oseltamivir and GS4071 map along the same line given that they have the same number of polar atoms (N + O).

8.2 PANORAMIC VIEW OF HIV-LAND: HIV-1 PROTEASE AND REVERSE TRANSCRIPTASE

The human immunodeficiency virus (HIV) pathogen does not need any introduction and neither do its two most heavily studied, analyzed, and explored enzymatic targets: HIV-1 protease (HIVP) and HIV-1 reverse transcriptase (HIVRT). In the context of the AtlasCBS representation, the area of CBS related to these two critical targets of HIV-1 therapy is immense. These examples reflect the uneven sampling of CBS stored within the SAR databases: thousands of measurements for a few targets and very sparse sampling of the vast majority of other targets of biological (not necessarily biomedical) interest.

Figure 8.3A presents an AtlasCBS window displaying the area of CBS encompassed by both targets. Compounds active against the protease are colored in red and in blue the ones with some measured activity toward the RT. The data were extracted by entering into the corresponding window the PDB codes for the crystal structures of two different target−inhibitor pairs: 1OHR (HIVP-Nelfinavir) and 1FK9 (HIVRT-Efavirenz). Other marker compounds were highlighted, in particular the two corresponding to the two extremes. The smallest and least polar compound (rightmost) is shown in the **Selection** window as active against the RT ($pIC_{50} \sim 2.45$), and the one furthest to the left, p2/NC, a substrate analog peptide inhibitor containing $NPOL = 19$. This wedge of CBS contains 2,774 compounds: 1,346 for the protease and 1,428 for the RT and includes only data for the wild-

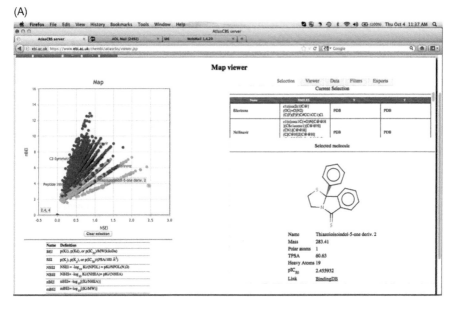

Figure 8.3 (A) The wide area of CBS occupied by compounds active against HIV targets. AtlasCBS window obtained as response to entering PDB codes 1OHR (Nelfinavir) and 1FK9 (Efavirenz). Colors were reset with the set color option. The wide segment of the chemical space corresponds to compounds active against HIVP and HIVRT. The small compound displayed is the rightmost in the map with activity against the HIVRT enzyme. Peptide p2/NC, a substrate analog, is at the other extreme. (B) Close-up of the region of drugs against HIVP. Detail of Fig. 8.3A where the majority of the effective HIVP compounds map. The yellow arrows outline a virtual trajectory (not traced by the AtlasCBS software) from the earlier to the latest compounds increasing in efficiency and improving their physicochemical properties. Further to the right still map the di-cyclic urea compounds that did not make to the clinic. Tipranavir "analog 7" (centered dot above) marks one of the many compounds synthesized along the discovery path to the final drug. (C) Tipranavir optimizes NSEI and nBEI. Export screen from AtlasCBS zoomed in to show the details of the chemical compounds near Tipranavir and Nelfinavir. The compounds line up along three different NPOL lines (NPOL = 6, 7, 8). Tipranavir analog 7 can be taken as a reference point to relate to Fig. 8.3B.

type sequence of corresponding targets. Other entries are available for a wide variety of mutants for both enzymes that were explored and studied to address the issues of resistance and specificity. It is not practical, within the limitations imposed by a server application, to explore these data in detail but a few insights can be easily extracted from this graphical representation.

It is evident from this representation that the compounds directed against the HIVP are considerably more polar than the ones against the RT; however, the two wedges are not well separated. A glimpse of the slice corresponding only to the area of CBS active against the HIVP has been presented in a previous publication[3] and also in Part II (Fig. 5.2). The variety of the chemical entities active against the

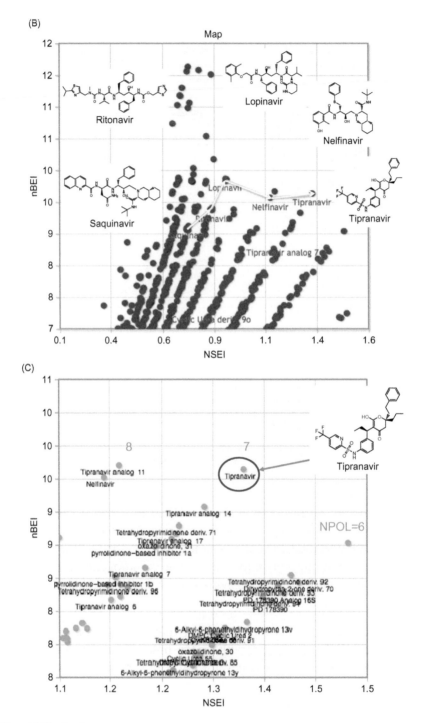

Figure 8.3 (Continued)

protease target alone can be appreciated from Fig. 5.2. The span of the chemical space covers compounds with a polarity ranging from a hydroxyl derivative of fullerene to a dipeptide containing 34 polar atoms (DP). The size efficiency as described by nBEI ranges from 5 to 14. It can also be appreciated that some of the earlier HIVP inhibitors in the clinic (NPOL > 11, Saquinavir, S) violated three of Lipinski's guidelines (red).

Two more insights can be gained from the rapid inspection of the AtlasCBS windows displaying the HIVP data alone (Fig. 8.3B). First, there is distinct migration in the NSEI−nBEI efficiency plane of the compounds against the protease from the early, peptide-mimetic inhibitors like Saquinavir and even the more potent and efficient Ritonavir and Lopinavir, toward the more drug-like Nelfinavir and finally Tipranavir (yellow arrows). Similar trajectories can be drawn in different efficiency planes. Second, the final compounds tend to occupy the upper-most position of the lines, indicating that the most efficient compounds probably represent the best compounds for their series (Fig. 8.3C). This point has been discussed and illustrated in more detail in previous publications[3−5].

The area of CBS occupied by the compounds active against the HIVRT is also rather large although as indicated corresponds to less polar compounds. Zidovudine (Retrovir[TM] or AZT), the first effective drug against HIV, was approved in 1987. It was a nucleoside analog first synthesized in 1964 as an anticancer agent (Fig. 8.4A). Members of this class are relatively polar and bind in the active site of the enzyme. HIV has a very rapid replication rate and since it lacks proofreading enzymes for the DNA copies generated by the RT, resistance against AZT-like compounds develops very rapidly. Non-nucleoside reverse transcriptase inhibitors (NNRTIs), such as Nevirapine, were the second generation of RT inhibitors. These compounds do not bind at the active site but rather at a second site, referred to as the allosteric NNRTI pocket. Binding by Nevirapine induces a conformational change that renders the enzyme inactive. Nevirapine and the later NNRTI compound Efavirenz have been studied extensively and structures of these compounds and many other NNRTI are available from the PDB for native structures, as well as the mutants at positions Leu100, Tyr181, and Tyr188 that are critical for resistance (Fig. 8.4B). The extensive data on compounds active against the HIVRT available

(A)

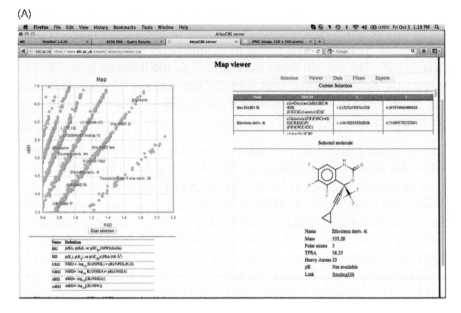

Figure 8.4 *Exploring the HIVRT territory using AtlasCBS. Different views of the portion of the AtlasCBS map related to compounds active against the HIVRT. (A) The Viewer/Selection window with the structure of Efavirenz 4i analog. (B) Close-up, export window, from the server showing different series, that is, thio-DABO and Chloro-TIBO. The shift in CBS from Nevirapine to Efavirenz (less polar and more efficient) is indicated by the red arrow. (C) Affinity data from BindingDB for wild-type (blue) and three mutants: ocher, Leu100Ile; red, double Leu100Ile/Lys103Asn; and green, Tyr188Cys. Note how the position of the compounds (e.g., Efavirenz) along the line (same slope: red arrow) changes depending upon the affinity: efficiency for the mutants is typically lower than for the wild-type enzyme, unless they were designed to overcome resistance for that mutant.*

from BindingDB can be conveniently explored by entering the following access codes: 1FK9 (Efavirenz), 3HVT (Nevirapine), 1JLG (UC-781, L100I mutant), 1S1V (TNK-651), and others. Upon entering any of these, the server will extract all the available affinity data against the RT target from the database and highlight the position of the complex in the NSEI−nBEI efficiency plane. Entering in the search box "Efavirenz" will extract all of the entries containing this word and their positions in the map can be followed by selecting them. An example is presented in Fig. 8.4C. The reader is encouraged to explore these data by this strategy or by selecting the corresponding target on the tabs for the databases.

8.3 A SNAPSHOT FROM VIRTUAL LAND: MAPPING CALCULATED EFFICIENCIES

Probably the most important component for an effective and efficient drug-discovery effort in the future would be the possibility of

(B) Map

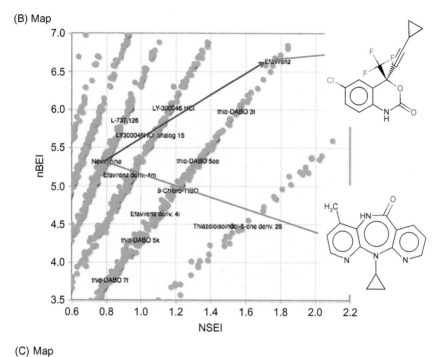

(C) Map

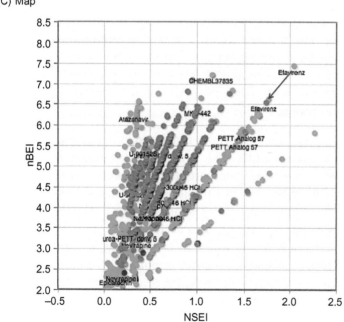

Figure 8.4 (Continued)

calculating affinities between targets and ligands with a certain degree of accuracy and confidence, given a well-defined 3D structure of the target[5]. The assumption that this is judiciously possible today is at the core of a multitude of projects in both industry and academia. There are indeed limitations to what the current methodology can do[6]. However, an accepted strategy to identify initial hits is to use computational approaches to limit the number of compounds experimentally tested from large libraries, or even from the limited collections used in fragment-based (FB) approaches. Docking calculations of ligands to targets are used routinely as a means of "virtually screening" collections of compounds ("chemical libraries"), as an initial step to identify possible hits that can later be tested experimentally. The cost-saving benefits of this approach are evident. The importance of reasonable theoretical estimates of the binding constants of proposed compounds is also critical to optimize and expedite drug design.

There have been significant developments in the field of computational chemistry and free energy calculations to improve affinity estimates. Nonetheless, robust ranking of compounds on the basis of their theoretical affinities or docking scores is still unsatisfactory. From the viewpoint of more effective virtual screening, an important development has been the concept of "enrichment" of libraries in the active compounds. This enrichment has been an effective way to provide tangible results in the form of a collection of compounds deserving experimental testing.

Along those lines, an important improvement was the notion of effective "decoy" compounds that could be used to benchmark the various software tools and compare their respective strengths and weaknesses. A rather significant milestone was achieved by the establishment of the "Directory of Useful Decoys" (DUDs) at the University of California San Francisco (UCSF)[7]. The initial version contained 2,950 ligands for 40 different targets that in various combinations provided a database of 98,266 compounds. We used this initial DUD library in one example of virtual screening using the AtlasCBS but a more recent version (DUD-enhanced, DUD-E) has been released[8].

The target of interest was catechol-O-methyltransferase (COMT)(EC.2.1.1.6), an important enzyme involved in the methylation of catechol-related neurotransmitters (e.g., epinefrin, norepinefrin) in higher organisms, including man, using *S*-adenosyl-methionine (AdoMet) as the methyl carrier. There are two isomorphs (soluble, s-COMT, and membrane-bound, MB-COMT) originating from a single gene. Interest in this enzyme has increased recently due to its importance as a target to develop new drugs with activity in Parkinson's disease and other neurological disorders.

This example illustrates how it is possible to use theoretical estimates of the binding affinities (K_i) from virtual screening to map the results in efficiency space. To this end, the affinities obtained for the known binders and for a reasonable collection of decoys assembled from the DUD library will be combined in the same Cartesian plane.

The collection of 430 decoys from DUD was augmented with 11 known COMT binders[7]. The SMILES strings were used to generate the 3D structures. This was followed by conformational analysis of the ligands and assignment of partial charges calculated by the standard methodology. Docking calculations were done using the CRDOCK software and the free energy of binding was estimated by the nonbinding terms of the AMBER99 force field[9]. The results were prepared in a Comma separated value (CSV) formatted file accepted by AtlasCBS as indicated in Part I. Briefly, a file consisting of a list of compound name, SMILES, affinity type (i.e., K_i) and value in nanomolar units for each compound was prepared. This file was uploaded into the AtlasCBS "UserSet" using a private user login (Part I) and the resulting screen is presented in Fig. 8.5A.

The striking result of this simple exercise is that the compounds that occupy the top of the three lines (16477, 16474, 16483), corresponding to NPOL=2,5,10 (respectively) are in fact three known COMT inhibitors included in the chemical library. This result provides independent confirmation of the docking protocol and of the quality (in terms of efficiency) of the compounds. This would suggest that promising hits for further experimental testing or study should be selected from the ones immediately below the known binders or, if a change in the number of polar atoms is needed, select the top compounds from the other lines (Fig. 8.5B).

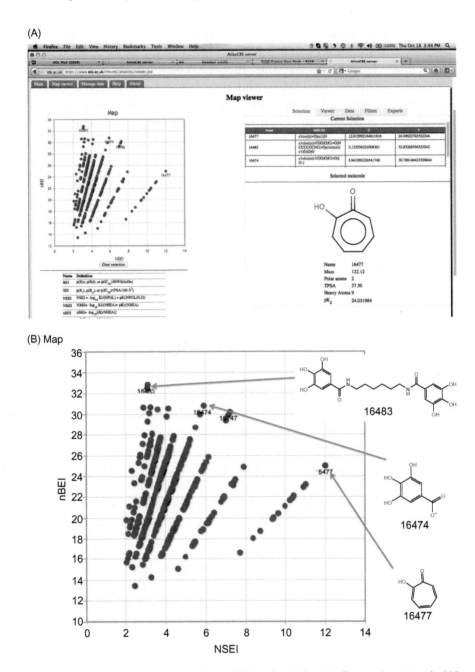

Figure 8.5 View of the virtual screening results for COMT in the NSEI–nBEI efficiency plane. (A) AtlasCBS map window after uploading the "UserSet". The **Selection** window shows compound number 16477. (B) Export window annotated with the structures of three of the known binders that map at the top of the corresponding NPOL = 2,5,10 lines.

8.4 WETLANDS OF CANCER—KINASE INHIBITORS (Gleevec, Iressa)

The expanse of CBS related to cancer is enormous. The biological processes connected in one way or another to initiation, growth, and development of cancer are multiple. Consequently, the number of macromolecular targets that have been explored to identify, design, and develop effective therapies directed toward the various forms of cancer is also large. The marketed compounds range from the early DNA-binding, platinum-containing drug (Cisplatin, Platinol^{TM}) of the late 1970s, to the newest breast cancer therapies related to the inhibition of enzymes involved in the biosynthesis of estrogen, known as selective estrogen receptor modulators (SERMs). Protein kinases, the enzymes that regulate a myriad of biological processes by phosphorylating other proteins, have also been the object of intense exploration. The completion of the human genome revealed that there were 518 different kinases[10] and the concept of "kinome" was followed by others "omes" (i.e., proteome, metabolome, reactome) to indicate maps of biological complexity in target and ligand space. In cancer therapy, a great deal of effort has been directed toward the development of potent and selective inhibitors of one (or just a few) of these enzymes.

Among them, inhibitors of the epidermal growth factor receptor tyrosine kinase (EGFR-TK) have been the focus of intense medicinal chemistry efforts. Already in the 2006 BindingDB database (Fig. 1.2), there were nearly 1,000 affinity entries for this target. Currently, AtlasCBS has access to over 1,800 entries. These data can be extracted from the database by selecting the EGFRTK target or using the PDB entry 2ITO (EGFR mutant–Gefitinib complex). A section of this map is presented in Fig. 8.6 with the **Selection** window showing the structure of Staurosporine, a broad-spectrum, tyrosine kinase inhibitor, isolated in 1977 from *Streptomyces staurosporeus*. The extent of CBS occupied by the inhibitors of protein kinases can be appreciated by the thickness of the lines that follow the NPOL slopes in the NSEI–nBEI plane (Fig. 8.7, upper left inset), particularly at the lower NPOL numbers. The spread of the lines diminishes at the higher values. The figure also illustrates the position in the map of some of the earlier, simpler, kinase inhibitors (Erbstatin, Genistein). The more recent and better-known inhibitors such as Erlotinib (Tarceva^{TM}) and Gefitinib (Iressa^{TM}) are also shown with different values of IC$_{50}$, derived from

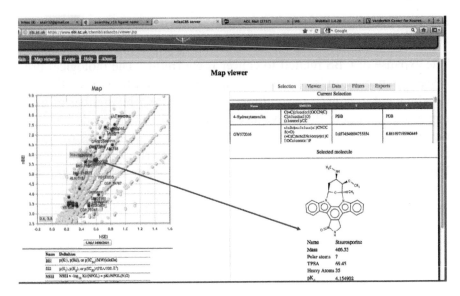

Figure 8.6 Efficiency plane of compounds active against EGFR tyrosine kinase. Mapping of the available data for the human form (1,823, IC$_{50}$) within BindingDB. The broad-spectrum inhibitor Staurosporine is annotated.

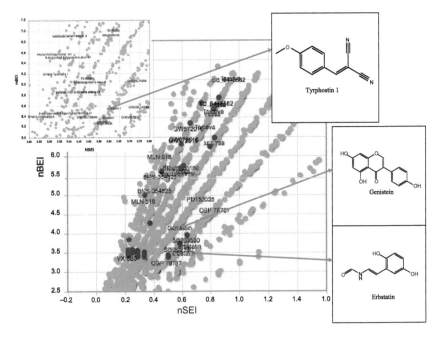

Figure 8.7 Export window of the data for EGF receptor kinase. Window of EGFR-TK inhibitors extracted from AtlasCBS: from the natural product Erbstatin (below) to the more recent Tarceva^{TM} and Iressa^{TM}. Structures of three natural inhibitors Erbstatin (bottom), Genistein (middle), and Tyrphostin 1 (top) are annotated. Inset. Close-up of a small region of the window showing how populated the chemical space is at the lower NPOL values.

different assays. Many compounds that were not designed to inhibit the EGFR kinase are also presented in the map because they have been tested for activity against this milestone tyrosine kinase.

With the number of targets for the different types of cancer being so numerous, I have decided to illustrate another one separately because of its importance in the human population. Breast cancer affects one in every eight women, often with immense physical and emotional consequences. One macromolecular target for therapy against this condition is the estrogen receptor (ERα) and Tamoxifen (Nolvadex[TM]) was an early selective modulator of the action of the estrogen-regulated transcription factor. Figure 8.8 shows the compounds extracted by the AtlasCBS server upon entering the PDB code 3ERT corresponding to 4-hydroxytamoxifen (186 compounds), the active metabolite of Tamoxifen, which acts as an ER antagonist in breast tissue, blocking the binding of estrogen to the ERα with certain selectivity. The position of other active agents of interest, for example, Raloxifene (Evista[TM]) can be found by entering the full or partial name in the **Selection** window. Raloxifene is a second-generation

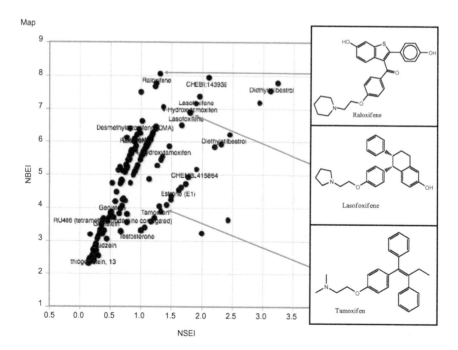

Figure 8.8 *Efficiency plane of compounds against the ERα. Data extracted from BindingDB with several landmarks compounds and drugs annotated: Tamoxifen, Lasofoxifene, and Raloxifene.*

SERM initially used to prevent osteoporosis in postmenopausal women, and later also employed for reducing the risk of invasive breast cancer in postmenopausal women. It is more specific for the ERβ with effect on bone and cholesterol metabolism but behaves as a complete estrogen antagonist on mammary gland and uterine tissue (Drugbank)[11].

In recent years, another target of considerable interest for certain types of cancer has been BCR−ABL, the protein encoded by the anomalous fusion of BCR and ABL genes. This fusion results in the expression of a deregulated tyrosine kinase that remains in the active form. The combined BCR−ABL was soon linked to chronic myelogenous leukemia, and the agent Imatinib (Gleevec[TM]) was introduced in 2001 as a potent tyrosine kinase inhibitor with therapeutic action[12]. In addition to Imatinib, BindingDB has affinity values (K_d) for other landmark kinases under the ABL (*H. sapiens*) target list. They are presented in Fig. 8.9. Note the presence of additional points for each target−ligand complex. This corresponds to the affinities for the Q252H mutant, showing that Imatinib has indeed a significantly lower affinity for the mutant. Comparing the affinity of the milestone compounds against various mutants is very important since they could become inactive upon mutation of the target enzyme BCR−ABL.

8.5 VISTAS FROM NEUROTRANSMISSION: ACETYLCHOLINESTERASE

Effective communication among neurons and between nerve and muscle cells is a critical component of the actions and movements that we observe in the animal kingdom of which we are a part. Neurotransmitters are typically small molecular entities that are released from nerve cells in the presynaptic gap and act as the receptors on the other side of the synapsis. Extremely rapid chemical events mediate these processes that depend on a wide variety of neurotransmitters such as glutamate, dopamine, serotonin, acetylcholine, and many others. Of critical importance for nerve−muscle communication are choline and its ester derivative acetylcholine. The enzyme choline acetyltransferase synthesizes the ester that is the effector on the muscle, and acetylcholinesterase (AchE) breaks the ester and recycles back the choline for the next contraction.

In 1991, researchers from the Weizmann institute unveiled the first 3D structure of AchE from the Pacific electric ray (*Torpedo californica*)

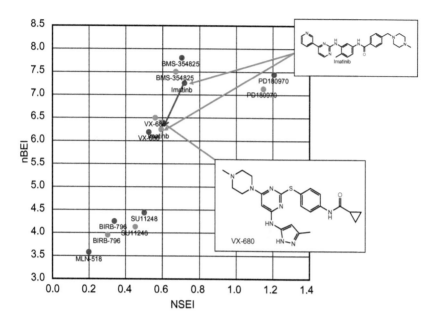

Figure 8.9 Efficiency differences of BCR–ABL inhibitors. Several milestone compounds of protein kinase therapy are mapped, showing their efficiency values against BCR–ABL wild-type (red dots) and the Q252H (green dots) mutant. Other mutants can be extracted from BindingDB and compared, for instance, the T315N mutation.

opening an astonishing window into the atomic mechanisms of this exquisitely specific and rapid enzyme[13]. Being so critical for the well-being of organisms, AchE has been the target of extremely powerful poisons (i.e., organophosphorous) that act by inhibiting its action irreversibly. Nowadays, AchE has come to the forefront of human interest again due to the cholinergic hypothesis of Alzheimer's disease (AD), which suggests that at least some of the cognitive decline observed in AD patients is due to low levels of acetylcholine. Inhibition of AchE would be an effective way to raise acetylcholine levels. This approach does not "cure" AD but rather attempts to palliate some of the effects of the disease.

Both BindingDB and ChEMBL contain a very large number of entries relating to various forms of AchE (*T. californica,* human, rat, mouse, and others) with active molecular entities obtained from natural sources (i.e., Huperzine A) and also synthesized in pharmaceutical and academic laboratories. After the pioneering efforts of Sussman and colleagues[13], there are now over 100 AchE structures in the PDB, including a large number of AchE–ligand complexes (e.g., AchE–tacrine,

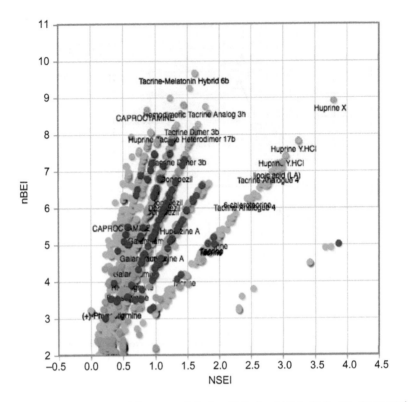

Figure 8.10 A glimpse at some of the compounds having affinity for AchE in BindingDB. Export window extracted from AtlasCBS—Blue and amber correspond to IC_{50} (1,105) and K_i (234) affinity values available for the human enzyme; red depicts the values for mouse IC_{50} (104). Huprine X, the compound resulting from "cutting and pasting" elements of Huperzine A and Tacrine, has superior efficiency in size and polarity than the corresponding separate molecules. Other compounds, such as the natural product Galanthamine and the AD drug Donepezil (Aricept^{TM}), are also annotated in the efficiency map. Donepezil is repeated several times along the same line because there are IC_{50} and K_i values for the human and mouse enzymes. Note that analogs of Tacrine with the same number of polar atoms map along the same NPOL line (Fig. 8.11).

AchE–Galanthamine, and AchE–Donepezil (Aricept^{TM}): 1ACJ, 1DX6, and 1EVE, respectively).

The vast majority of the affinity data can be displayed in the AtlasCBS page by selecting the AchE target within the BindingDB database.

Figure 8.10 shows the exported AtlasCBS NSEI–nBEI efficiency map with the data for the affinities measured for IC_{50} (blue) and K_i (ocher) for human, and IC_{50} (red) for mouse; other targets are also available. Several compounds of interest are annotated that can be followed dynamically in the window and whose structures are illustrated in Fig. 8.11A,B. Huperzine A is a natural product used in traditional

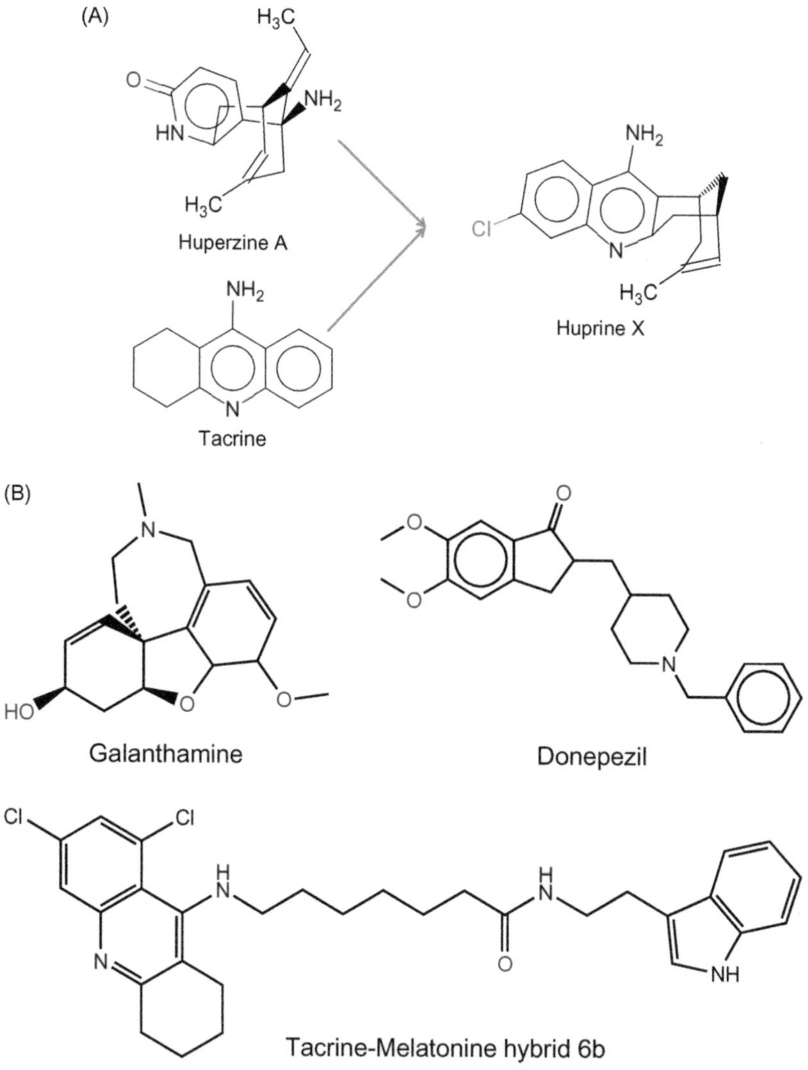

Figure 8.11 Compounds associated with inhibition of AchE. (A) Huprine X results from merging chemical moieties from Huperzine A and Tacrine. (B) Galanthamine (a natural product), the marketed drug Donepezil and a hybrid of Tacrine and Melatonine (hybrid 6b) are shown. These compounds are mapped in Fig. 8.10.

Chinese herbal medicine that inhibits AchE, and Tacrine (Cognex^{TM}) is an approved drug that also acts by the same mechanism. The combination of both drugs by "cutting-and-pasting" chemical fragments results in the hybrid molecule Huprine X that is a very potent ($K_i = 26$ pM) inhibitor with high efficiency in size and polarity. A different hybrid molecule (Tacrine–Melatonine hybrid 6b) achieves high affinity in size but does not have the balanced properties that Huprine X has (Fig. 8.11B). The effect of "line hopping" by having a different number of NPOL numbers is clearly illustrated. The user is encouraged to reproduce this window and to explore different compounds and regions of the map by selecting the corresponding portions with the mouse. Further details about the different compounds can be found by linking back to BindingDB, ChEMBL within the **Selection** window.

8.6 GLIMPSES OF RECEPTOR LAND: G PROTEIN-COUPLED RECEPTORS

The previous vignette introduced various neurotransmitters and discussed targets related to their degradation or processing. Here, I'll present an example of the biological molecules that responds to those effectors, namely, receptors that elicit a response upon binding to those unique molecules. The field of receptors in biology is immense given its tremendous importance in cell-to-cell communication. There are many types of receptors (adrenergic, dopaminergic, opioid, and others) in biological organisms, including the ones that bind and recognize the neurotransmitter acetylcholine that was previously discussed.

Traditionally, opium extracts from the plant *Papaver somniferum* have been used for analgesic or recreational purposes for thousands of years. Only very recently, however, have researches been able to unveil the mode of binding of these pharmaceutical agents to their receptors in atomic detail. The principal challenge has been to isolate the receptors in their native (active) state, because they are membrane-embedded proteins that denature (i.e., unfold) in an aqueous environment. Receptors targeted by the active substances related to opium (opioids) are members of the class of G protein-coupled receptors (GPCRs). The continued effort devoted to understanding the structure and function of the GPCR family of receptors, at the molecular level, has been awarded the Nobel Prize in Chemistry 2012 (Robert. J. Lefkowitz and Brian K. Kobilka). Early this year, a set of four papers published in *Nature*[14–17] has revealed structural details of the binding of therapeutic agents to opioid receptors that will undoubtedly have tremendous impact for the pharmacology and therapies related to these targets. This work also represents an amazing achievement for the structural biology community, since these receptors are extremely difficult to crystallize, and new methods had to be developed to capture these unique targets into well-diffracting crystals.

As illustrated in Part I (Fig. 3.2B), GPCR ligands occupy a very distinct part of CBS, distant from the more conventional targets. Typically, the structure of these receptors is based on a tight 7-helical-bundle domain (or transmembrane helices referred to as 7TM) embedded in the cell membrane. In BindingDB, a substantial amount of affinity data about these compounds is stored within "delta/kappa/mu opioid receptors", which are the prefix by which the three main classes are known: μ for (morphine, μ-OR), δ (for vas deferens, δ-OR), and κ (ketocyclazocine, κ-OR) receptors. The polypeptide chains for these three receptors have a high degree of homology ($>70\%$ identity) and the specificity for the ligands that each binds and the elicited biological actions are extremely subtle, based on the difference of a few amino acid residues in the binding pockets. The μ-, δ-, and κ-OR also respond to endogenous peptides secreted within the organisms: endorphins, enkephalins, and dynorphins among others. A latecomer to the opioid family of receptors is the nociception/orphanin FQ (N/OFQ) peptide receptor also known as NOP or ORL-1. This receptor is not as closely related as the other three ($\sim 60\%$ sequence identity) and has unique pharmacological properties. Various entries in BindingDB contain data for all four classes.

Figure 8.12 presents a window of the AtlasCBS for the four classes of opioid receptors mentioned above, showing the data (IC_{50}) available for three species: human (blue, 99), mouse (red, 7), and rat (green, 23), and a few (yellow) for which only K_i were available. Note, however, that the efficiency plane presented is not the typical NSEI−nBEI but rather SEI−mBEI. The difference in polarity among these compounds is more subtle than the NPOL atoms, therefore a different separation can be used in the angular coordinate to differentiate among them. In this plane, the separation in the angular coordinate is 0.01 PSA; thus small differences in the surface area can be used to discriminate among the compounds (see Table 2.1 and Appendix A).

Unfortunately, the earlier release of BindingDB did not include the affinity data for the recently published compounds; however, a few insights can still be extracted from the AtlasCBS window (Fig. 8.12). On the most polar region of the plane (near the blue vertical lines), one can see the Leu-Enkephalin, [D-Pen(2,5)]-enkephalin (DPPDE), and [D-pro10]Dynorphin A(1−11) (CHEMBL384584). These are some of

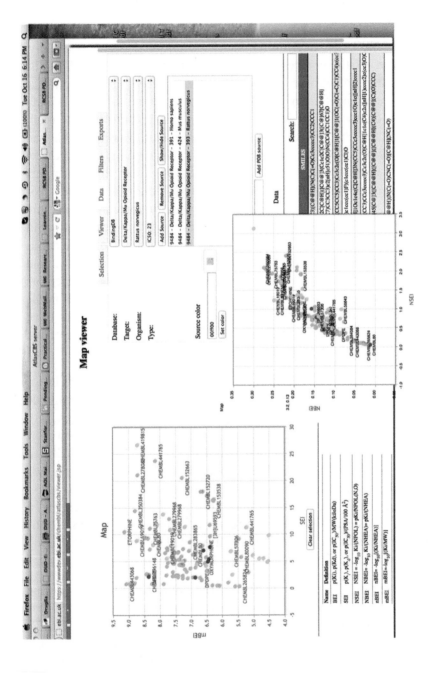

Figure 8.12 AtlasCBS window of opioid receptors (NSEI−mBEI). Available compounds against a combination of opioid receptors from various vertebrate species. Human: blue, mouse: red, and rat: green. Yellow was used to include other opioid-active compounds, for which only K_i data were available (see **Data** window). The variables for the x−y axes were changed using the **Viewer** tab and selecting a smaller section of the map (−5<SEI<30, 4.0<mBEI<9.5). Inset. Same data in the NSEI−NBEI efficiency map: notice the clustering in a very limited range of the angular coordinate NPOLINHA and the curving of the successive points toward higher polarity efficiency (x-axis).

the endogenous peptides that are active against these receptors, the latter one binding with high efficiency. A close-up of this map is presented in Fig. 8.13A and the structures of these peptides are shown in Fig. 8.13B.

A few important ligands are also highlighted in the efficiency plane and in the structure panels, in particular, Bremazocine[18] and the compound labeled CHEMBL19019 (Naltrexone) that is the marketed drug Revia (Depade™).

Finally, since the AtlasCBS is a window into the chemical and biological space, I would like to present the binding modes of the four agonists recently unveiled bound to the corresponding receptors: β-funaltrexamine (β-FNA), JDTic, naltrindole, and a peptide mimetic (Fig. 8.14A−D). They represent four different binding modes and views of the four different classes of opioid receptors (μ, κ, δ, and OFQ, respectively). It is my dream that in the future, the AtlasCBS tool would map each target−ligand pair in efficiency planes and also would allow the user to display the mode of binding, if structures were available, all at your fingertips.

The first one (β-funaltrexamine, PDB entry 4DKL) is a covalent inhibitor of the μ-opioid that is clearly seen (upper right) in the structure linked to the receptor by Lys233 (Fig. 8.14A). JDTic (4DJH) is a selective antagonist of the κ-OR (Fig.8.14B). Naltrindole (4EJ4) is a subtype-selective antagonist of the δ-opioid receptor (Fig. 8.14C). Finally, the structure of a peptide-mimetic inhibitor bound to the more distant ORL-1 is shown in Fig. 8.14D (4EA3). The impact of these structural results in the field of drug discovery is likely to be enormous.

8.7 FB LEAD GENERATION: LACTIC DEHYDROGENASE A

The concept of FB strategies in drug design was presented earlier in the historical context. These strategies encompass the general concept of fragment-based drug design (FBDD) and also more limited areas of the drug-discovery effort such as fragment-based lead generation (FBLG). FB strategies seek to identify small chemical entities (MW ≤ 200 Da), ideally at two or more proximal locations on the surface of the target, which can be linked together to create a larger, more potent and efficient compound. The widespread use of FB strategies gave the initial impetus to the acceptance of ligand efficiency (LE) as an adequate metric to rank fragments for further development[19].

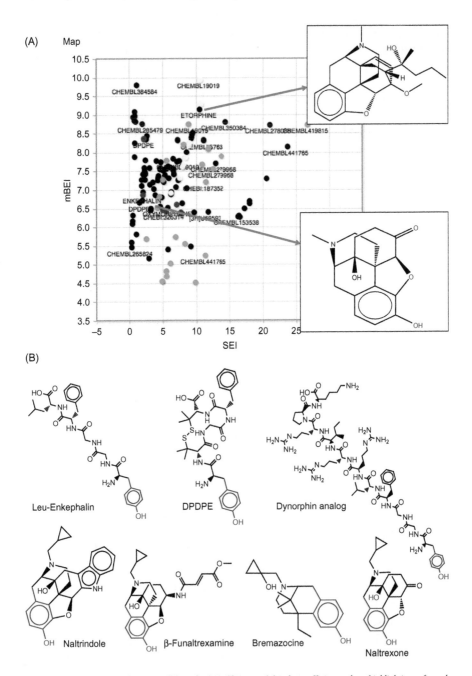

Figure 8.13 Bioactive peptides vs. small ligands. (A) Close-up of the above efficiency plane highlighting a few relevant compounds within a narrower range of data. Enkephalin and DPDPE denote the position of the two bioactive, naturally occurring, peptides. In addition, Oxymorphone and Ethorpine are active small-molecule agonists. CHEMBL350384 indicates the position of Bremazocine[18], a κ-OR agonist with potent analgesic and diverse pharmacological properties. (B) The chemical structures of some of compounds mapped in Fig.8.13A. The progression from the large, polar, bioactive peptides to smaller compounds with balanced physicochemical properties is clear, as reflected in the inset of Fig. 8.12. The red hydroxyphenyl ring highlights the active tyrosine pharmacophore.

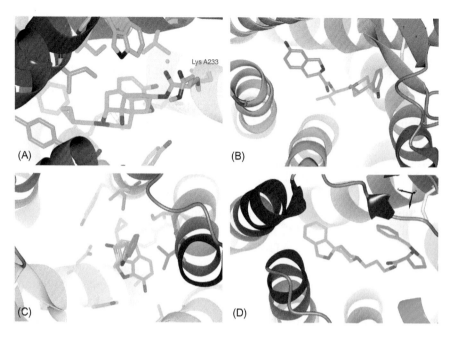

Figure 8.14 Views of the binding mode of the four different types of opioid receptors. The different transmembrane helices making the binding pocket can be seen decorated with the side chains residues in (A) and (C). (A) β-FNA covalently bound to μ-opioid via Lys A233. (B) κ-OR with JDTic inhibitor. (C) δ-OR with Naltrindole. (D) ORL-1 with peptide mimetic. The coordinates of the corresponding PDB entries were downloaded and the images created with the molecular graphics display program QtMG included in the CCP4 crystallographic suite. Ligands are colored by atom type (carbon: green, oxygen: red, and nitrogen: blue); small red spheres correspond to water molecules (See text for further details).

In general, FB strategies require a significant commitment of biochemical, enzymatic, and structural resources to be successful. In this section, I would like to use the ligand efficiency index (LEI) framework to analyze in detail the efforts of Ward and coworkers[20] describing the design and synthesis of novel inhibitors of lactate dehydrogenase A (LDHA) to identify chemical compounds that could validate the target in cell assays. This work illustrates and documents the FB approach at its best, considering all the possible issues and difficulties, from enzymatic assays to crystal soaking, and the considerable number of resources and personnel devoted to it. The authors consider the LE of their compounds throughout the paper but it is not clear that this parameter was particularly useful in their strategy, nor that the linked compounds had superior LE values than the initial fragments. The analysis presented and illustrated here will be more encompassing in that it offers a graphical representation in efficiency planes of the lead generation project and in doing so it uses two efficiency planes (NSEI and nBEI or NSEI and mBEI) found within the AtlasCBS toolset.

A brief review of the structural features of the LDH target will clarify the FB approach followed by Ward and colleagues[20]. LDH is an enzyme with a long structural tradition. It forms a tetrameric arrangement containing four chains encoded by two different genes, A and B. The combination of four subunits in all possible arrangements results in five isoforms (B_4, B_3A_1, B_2A_2, B_1A_3, and A_4). The crystal structure of the dog-fish muscle M_4 isoform was solved in 1970 and unveiled several novel structural concepts[21]. Of particular importance was its modular architecture: the structure consists of two functional domains. At the N-terminus, duplication of a classic Rossmann fold ($\beta - \alpha - \beta - \alpha - \beta$) provides the pocket for the NAD cofactor (NAD-binding domain) at the carboxy end of six parallel β-strands. A different but intimately connected second domain provides the substrate pocket. This modular structure has since been found in a multitude of complex enzymes. It is precise by finding and optimizing fragments for each of these pockets that Ward and colleagues[20] intended to find an optimized lead that would permit target validation in cell assays.

At the time that this work was published, the data were not available at BindingDB and I considered preparing an external file using the required information to upload the 40 or so compounds into my private UserSet of the AtlasCBS. Fortunately, Prof. M. Gilson, Director of BindingDB, offered to incorporate the data from the article to the database and make it accessible for my analysis in a timely manner. When the data were available, it was downloaded from BindingDB as a structure data file (sdf)-formatted file and kindly processed by A. Cortés-Cabrera into the required CSV file (using a Java script that is available upon request) to upload into my "UserSet", as has been indicated earlier. The final dataset containing 39 (red) compounds with K_d affinity values were measured by nuclear magnetic resonance (NMR) or BIAcore assays and 34 of them with IC_{50} measurements (green) from the enzymatic assay (73 total). The data are summarized in the published article for compounds numbering 1−34 and the compounds can be classified into six groups: (i) compounds from published LDHA inhibitors (1−8); (ii) compounds identified from fragment screens (9−14); (iii) compounds binding at the adenine pocket derived from 12 (15−18); (iv) compounds derived from FB screening against the nicotinamide/substrate pocket (20−21); (v) adenine and substrate pocket libraries (22–25); and (vi) linked compounds (26–34) (Classes I–VI, respectively). The structure of these compounds is presented in Fig. 8.15A–F clustered in the appropriate classes indicated above.

The overall plot of the compounds for which affinity data were updated in BindingDB for LDHA is shown in the NSEI–nBEI efficiency plane using the AtlasCBS tool in Fig. 8.16. It is worth

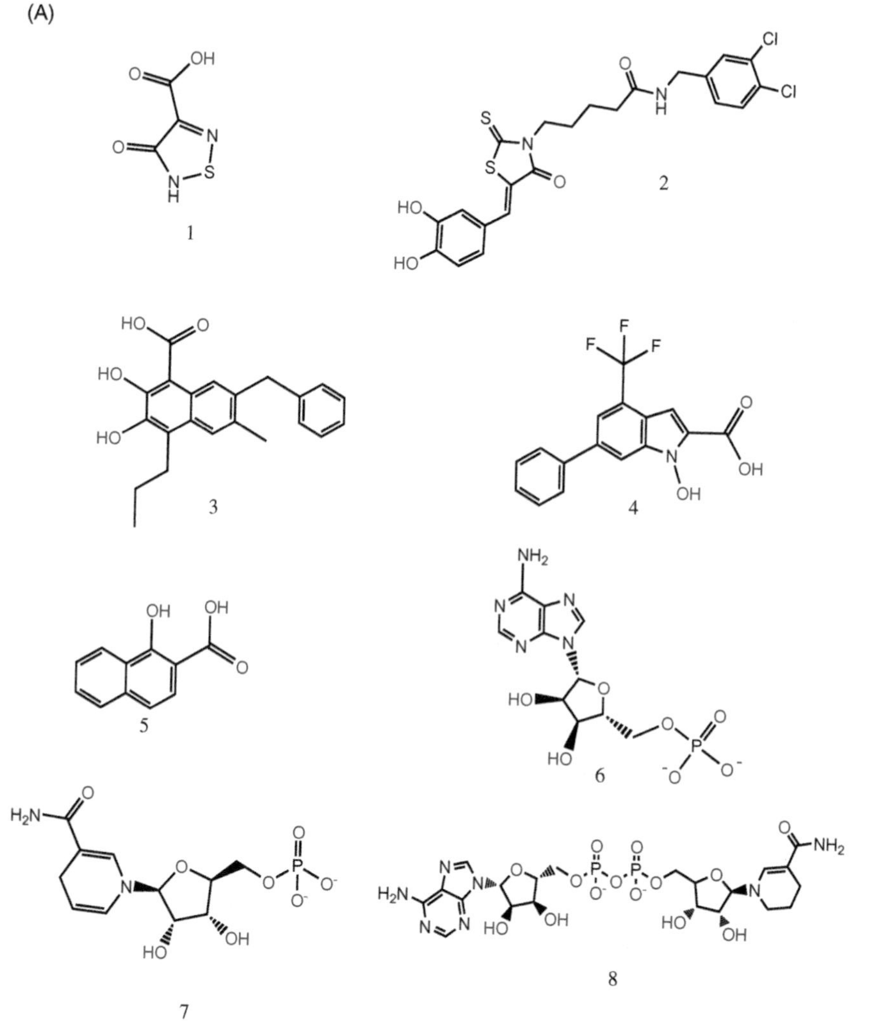

(A)

Figure 8.15 Compounds explored in FBDD for LDHA. (A) Class I. Proposed LDHA inhibitors (1–8). (A–F) Chemical structures and classes of compounds tested for LDHA. Compounds characterized in a FBLG for LDHA. Class I–VI correspond to parts (A–F). Adapted from tables 1–6[20].

(B)

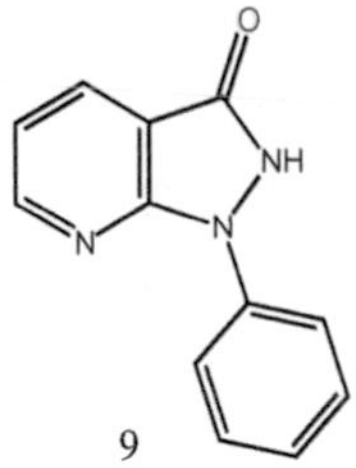

9

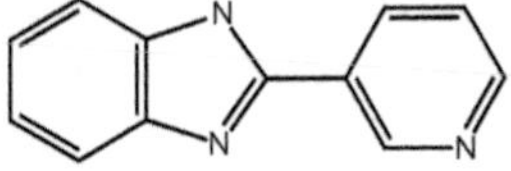

10

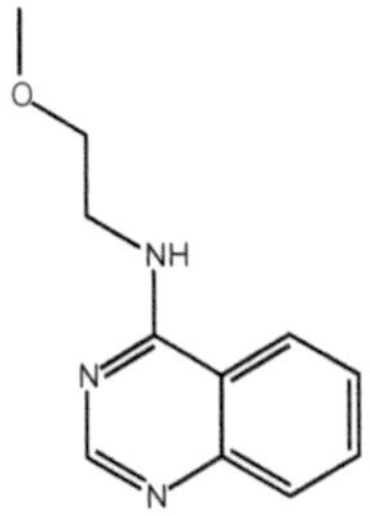

11

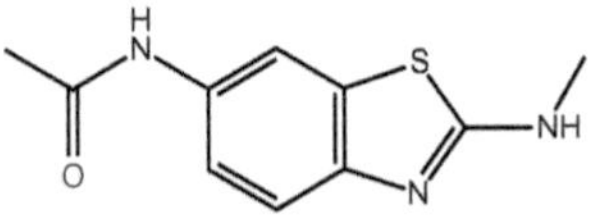

12

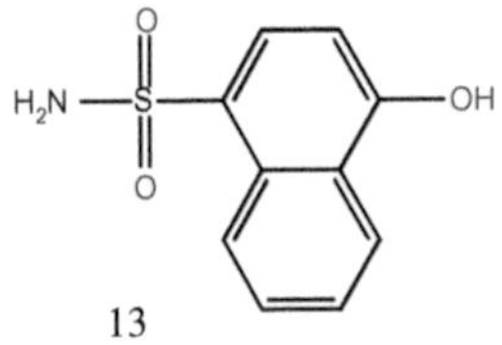

13

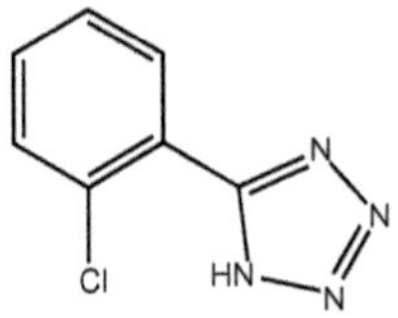

14

(C)

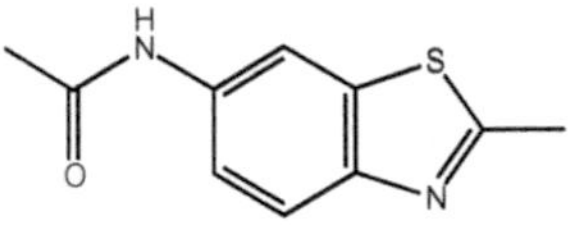

15

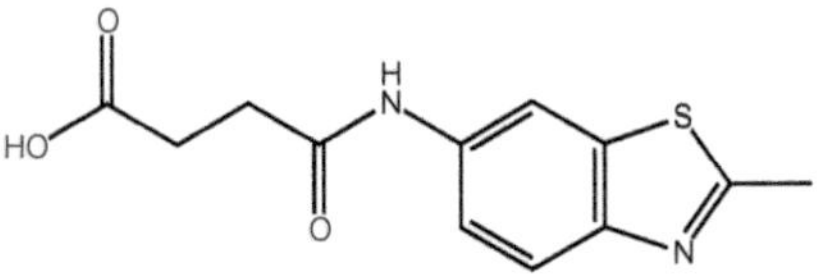

16

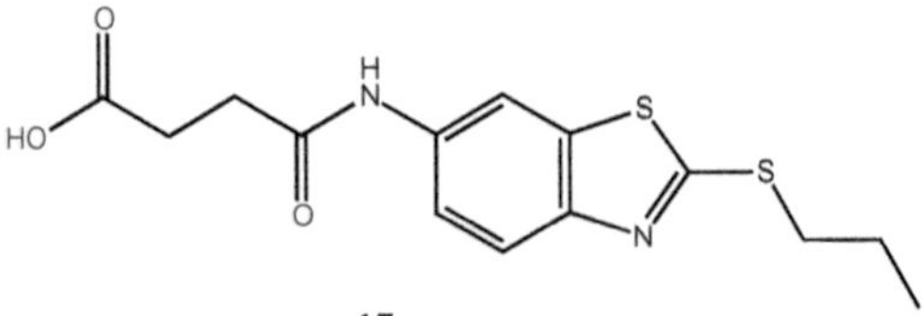

17

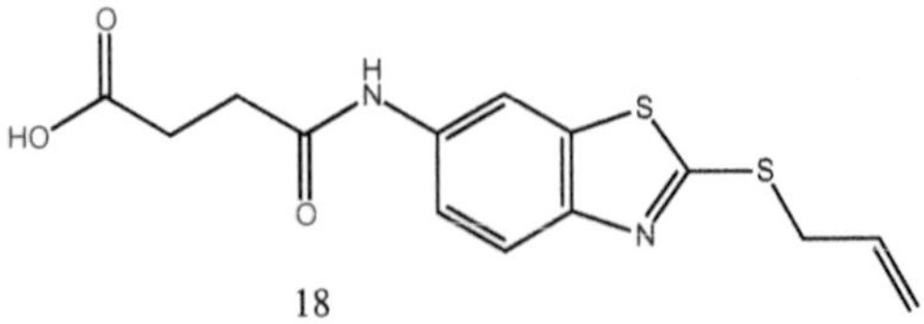

18

Figure 8.15 (B) Class II. Identified by fragment screens (9–14). (C) Class III. Bound to adenine pocket, derived from 12 (15–18). Adapted from tables 1–6[20].

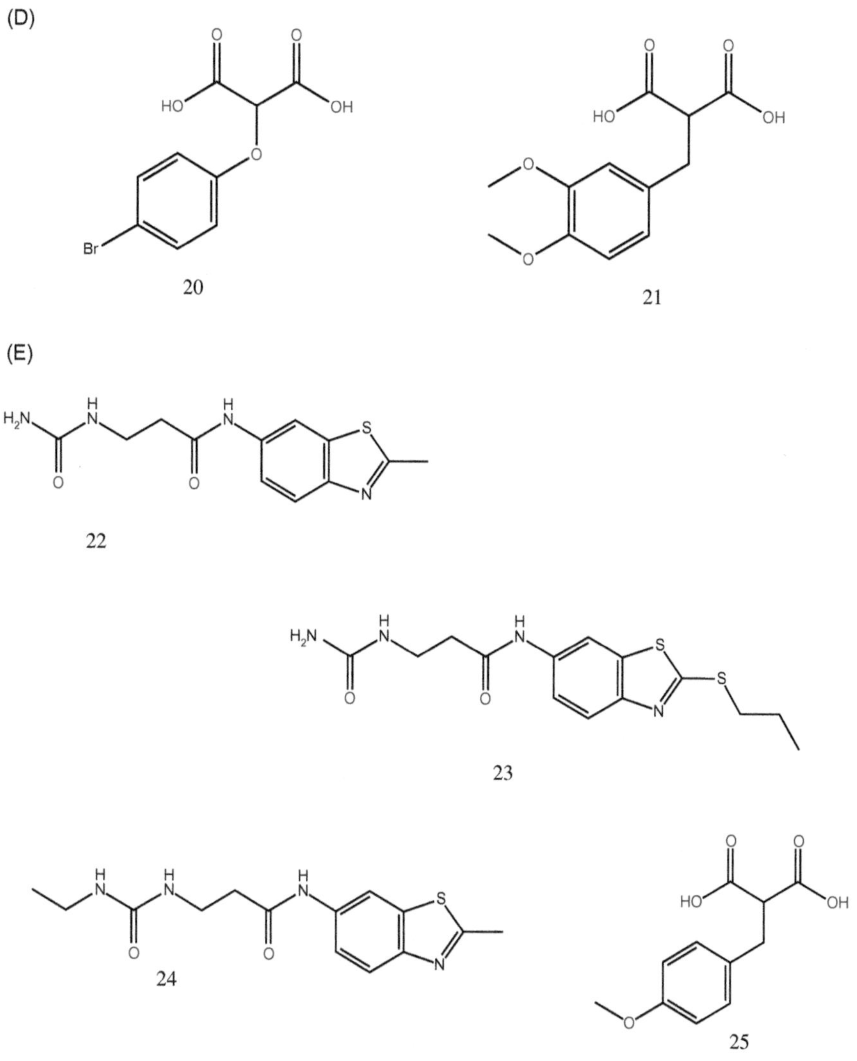

(D)

20

21

(E)

22

23

24

25

Figure 8.15 (D) Class IV. FB screening against nicotinamide/substrate (20–21). (E) Class V. Adenine and sub-strate pocket libraries (22–25). Adapted from tables 1–6[20].

Figure 8.15 (F) Class VI. Linked compounds (26–34). Adapted from tables 1–6[20].

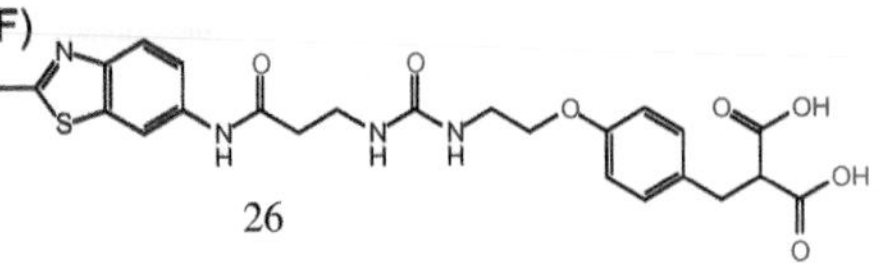

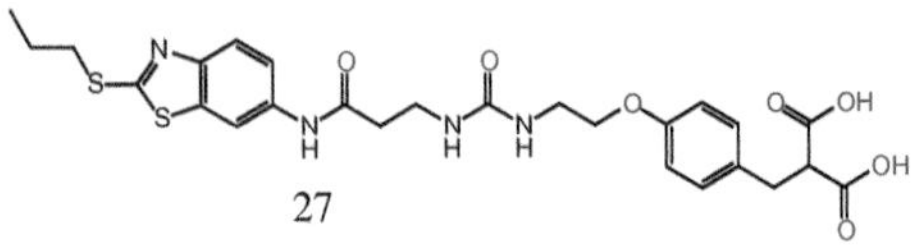

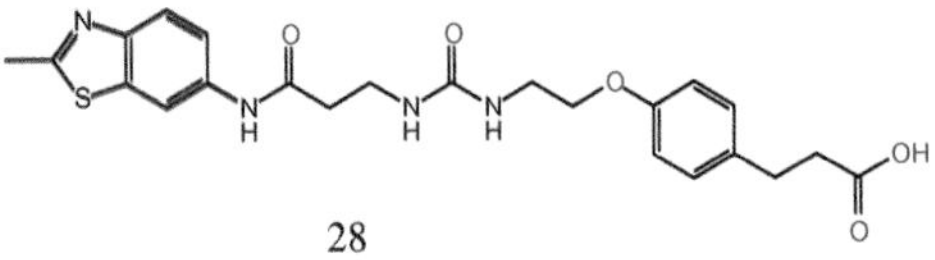

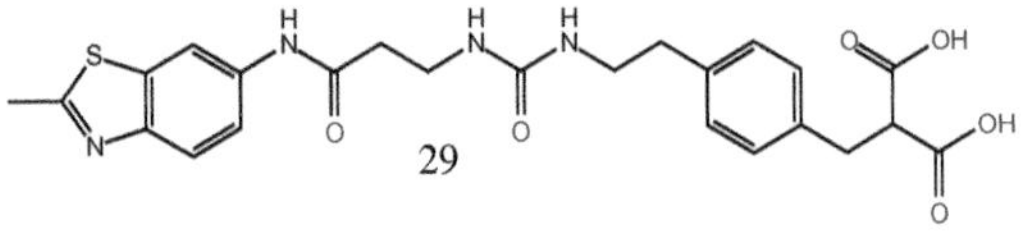

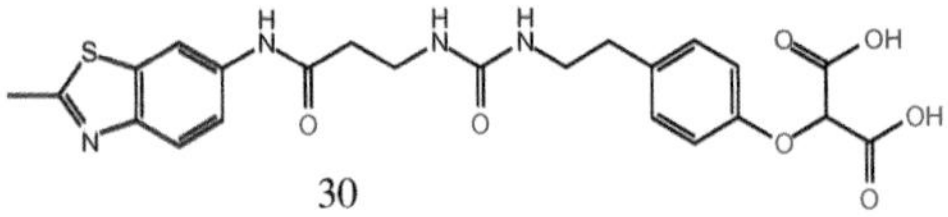

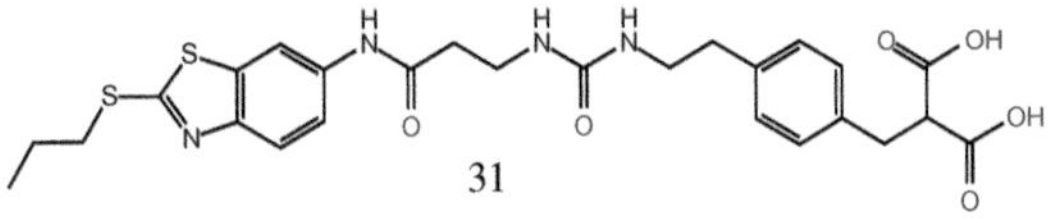

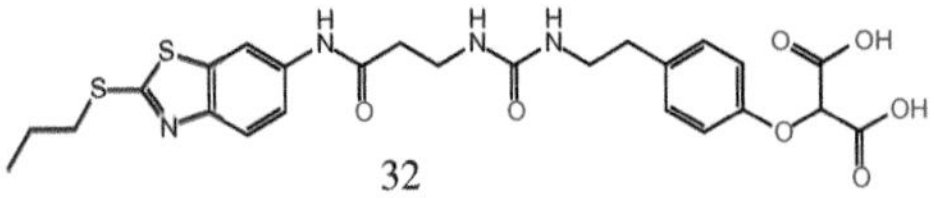

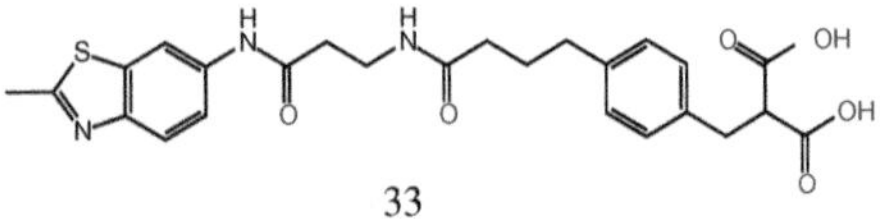

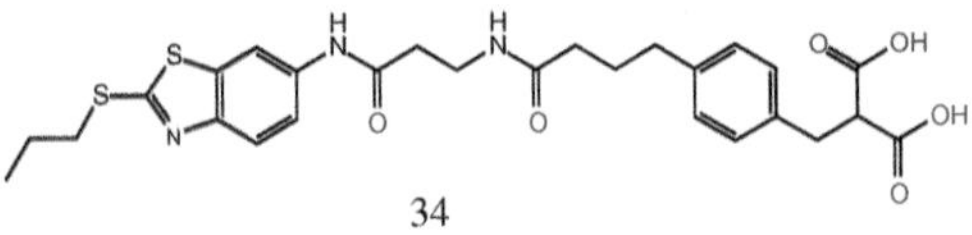

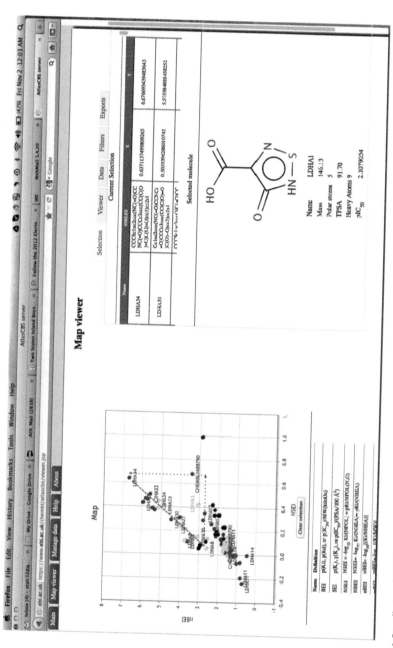

Figure 8.16 *Overall view of LDHA compounds considered for FBLG. Overall view of the data for the lactate dehydrogenase target in BindingDB. The plot was obtained by uploading all data for lactate dehydrogenase extracted from BindingDB. The data were downloaded first as a "structure data file" (sdf) from their server and then converted to a CSV (semicolon separated value, see Part I for details) using a Java script. Compounds with labels CHEMBL are not discussed because they do not belong to the FB project. The upper left map was critical for the project and the most efficient for the series LDHA34. The dotted components of the blue vector indicate the efficiency gains in 'polarity efficiency' (x, red) and 'size efficiency' (y, green) components.*

examining the general features of this efficiency map before going into the specific classes of compounds. Points correspond to the uploaded data extracted from the online BindingDB server and later uploaded as an external userset (K_d (red) and IC_{50} (green) measurements) (Fig. 8.16).

The efficiency plane shows most of the compounds presented in the previous classes with some of them being present more than once (e.g., LDH8) because the affinities were measured using more than one method (BIAcore: 12.8 μM, IC_{50}: 132 μM, and 9.4 μM for LDHA and LDHB). As discussed by Ward and colleagues[20], LDH8 was critical in providing a milestone for the goals of the project. The lower right panel shows compound 1 as the most efficient of all the published LDHA inhibitors (Class I).

It is apparent that the compounds cluster along two separate narrow wedges, related to the number of polar atoms included in their molecular structure. The fragment-like, small compounds in Classes I–V (except LDHA8), which contain a relatively small number of polar atoms (~10 or less), map along the slice with a lower slope. In contrast, the "linked" compounds with $NPOL(N + O) > 10$ atoms (Class VI) map along the steeper line that reaches the maximum efficiency for compound LDHA34 (red series). This is the linked compound that was selected for target validation in the cellular assays.

A close-up of the positions of the smaller compounds in the same efficiency plane is presented in Fig. 8.17A, obtained with the export option within AtlasCBS.

Within this limited region of the map, one can appreciate the high efficiency (in the IC_{50} assay) of compound 1, which binds to the substrate site tighter in the presence of NAD^{+20}. The relative efficiencies (in size and polarity) for compounds 7, 9, 11, 14, 17, 20, 21, and others can also be seen. In a real-life demonstration, it is possible to identify each single dot and display the 2D structure of the compound on the screen as shown in Fig. 8.16.

Focusing on the region of the plane where compounds resulting from linking the fragments map, which corresponds to Class VI using the NSEI−mBEI efficiency plane, one can see that for their respective assay conditions (K_d, red, LDH34* or enzyme IC_{50}, green, LDHA34****)

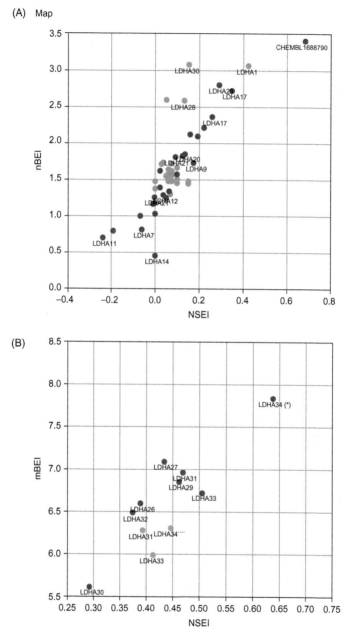

Figure 8.17 *Close-up of the two most significant sections of the overall efficiency map given in Fig. 8.16. (A) The smaller, less polar compounds corresponding to Classes I–V. (B) Region of the NSEI–mBEI efficiency plane corresponding to the larger, fragment-linked (Class VI) molecules. The linked compounds derived from joining the smaller fragments characterized for the adenine-binding pocket and the substrate/nicotinamide site. This region was extracted using the mouse within the AtlasCBS to select the corresponding values of the NSEI–mBEI plane. Figures exported from AtlasCBS with the* **Export** *tab. See Fig. 8.16 for a 'vectorial' summary of the optimization process.*

compound 34 was the most efficient of all the series. An imaginary trajectory line could be drawn in the map connecting the dots in their sequential numbers 26–34 (Class VI); note that compound 28 will be far below in the lower portion of the plane (Fig. 8.16A). The difference between the LE values for compounds 26 and 34, the initial and final of the linked compound series, is only 0.02 units (0.19 vs. 0.21)[20]. In contrast, using the combination of efficiency variables in polarity (NSEI) and size (mBEI or nBEI) provided within the AtlasCBS, it is clearly seen that the compounds represented by LDHA34 are far superior to any other molecules of the series in potency and efficiency, making them better candidates for further testing or development. A 'vectorial' summary of the gains from compound 8 to 34 is outlined in the efficiency map presented in Fig. 8.16.

In summary, using the dynamical representation of efficiency planes available at the AtlasCBS server it is possible to analyze and follow, graphically and quantitatively, the progression efforts along a FB strategy. It is also possible to compare the merits of different fragments or series and optimize the linked compounds toward an optimum compound with favorable physicochemical properties that will be, most likely, the most suitable for further target validation or subsequent studies.

8.8 A WINDOW INTO THE FUTURE—PROSPECTIVE USES OF AtlasCBS: TRANSTHYRETIN

An example of the prospective use of efficiency indices to guide drug discovery by Tanaka and colleages[22] was discussed in Part II. However, they did not use the fully developed variables that were introduced later, and thus I feel that I should complete these examples with one that makes prospective use of the unified framework suggested in this book, for which the methodology has been published recently[23].

Transthyretin (TTR) is a tetrameric protein that acts as a backup transporter for the thyroxine hormone (T4) in plasma. It is also the main carrier of vitamin A forming a complex with the retinol-binding protein. TTR has been implicated in various atypical amyloid diseases such as familial amyloid polyneuropathy (FAP) and familial amyloid cardiomyopathy (FAC), and certain degenerative conditions of the CNS such as senile systemic amyloidosis (SSA). Diflusinal, an anti-inflammatory agent related to aspirin, is a promising core structure for

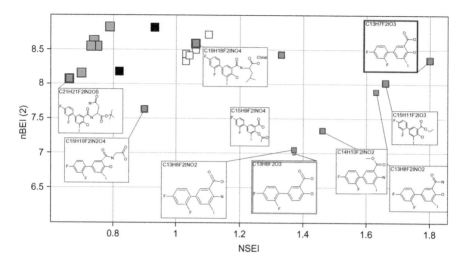

Figure 8.18 Retrospective mapping of diflunisal analogs in the NSEI–nBEI efficiency plane. Iodoflunisal (red box) is the most efficient of this series[29]. Blue box shows the decreased efficiency of the deionidated compound.

TTR-related amyloid diseases, and derivatives containing iodine atoms have shown some potential as lead compounds[24].

Blasi et al.[23] decided to use LEI-related variables with the goal of identifying more potent and efficient compounds by using the following strategy.

- First, represent in efficiency planes compounds related to diflusinal, with particular emphasis on the iodine-containing variants. A library of 40 salicylate derivatives (with and without iodine at position C5) was designed, prepared, and tested. The NSEI–nBEI efficiency plane for this new library was mapped and analyzed (Fig. 8.18). The plot clearly indicated that iododiflunisal (red box) was the most efficient of all the existing compounds.

- Second, using the iododiflunisal scaffold, a search was performed using the MMsINC database[25]. Commercially available compounds extracted from this database were merged with 500 internally proposed iododiflunisal analogs with various decorations in the aromatic rings. Thus, a final collection of 2,300 designed and commercial biphenyl compounds was considered.

- Third, the interactions between wild-type TTR and iododiflunisal (PDB entry 1Y1D) were analyzed using the software LigandScout[26] and a reasonable pharmacophore model was derived. Using this pharmacophore map, the 2,300 compounds were filtered extracting a

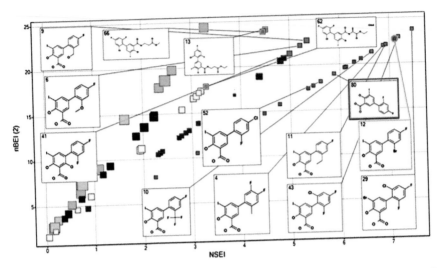

Figure 8.19 NSEI–nBEI efficiency plane of the 80 selected and docked compounds. Red squares: NPOL = 3; blue, yellow, black, light green, and cyan for NPOL = 4–8, respectively. Square size proportional to NPOL, increasing counterclockwise. Red box: iododiflunisal. Dark green: note the 12 compounds selected for synthesis and testing. Images prepared with Spotfire, since at the time AtlasCBS was not available. nBEI(2) is nBEI. Figures reprinted with permission from Blasi et al.[23] Copyright Molecular Informatics.

subset of 1,200 compounds that shared the same pattern of interactions as iododiflunisal.

- Fourth, the extracted subset was docked against the protein structure found in the TTR-iododiflunisal complex with the software MOE2009.10[27]. Using the MMFF94 force field and the appropriate protocols, it is possible to obtain an estimate of the binding free energy[23]. From the described methods, 80 compounds were prioritized by the best scores and from the calculated binding free energies it was possible to estimate the theoretical affinity constants K_i. These theoretical values were normalized to the experimental values obtained from calorimetry experiments[28] and were used to calculate LEIs for the 80 docked compounds. The resulting efficiency map is presented in Fig. 8.19.

Following this approach, it was possible to evaluate the relative efficiency of the various classes of compounds and follow the effect of adding or removing polar groups to the basic diflunisal scaffold. After visualization in the NSEI–nBEI plane, 12 promising compounds were selected, based on their positions in the map (Fig. 8.19), as they appeared to be superior to the reference compound. Ten of those compounds, theoretically more efficient than iododiflunisal, were actually synthesized and tested in the standard fribrillogenesis assays[24] and the results (unpublished data, Blasi et al., manuscript in preparation) validated the strategy. From the ten compounds tested (IC_{50}), four compounds showed similar or better inhibitory activity compared to the reference compound (iododiflunisal). In addition, using the experimental values it was possible to reevalute the earlier predictions (Fig. 8.19). From the ten analogs designed by using this prospective LEI approach, only three were predicted incorrectly; the rest of them were in good agreement with the predicted estimations. Moreover, the most promising compounds in the prospective LEI map were also the most efficient in the final evaluation. These results suggest that, used judiciously, it is possible to use the LEI framework prospectively to guide the lead optimization process and integrate the work flow with other computational methods such us pharmacophore modeling and docking. Further succesful examples of this methodology are needed to validate the approach, the variables and the selection criteria.

REFERENCES FOR PART III

1. Kim CU, Lew W, Williams MA, Liu H, Zhang L, Swaminathan S, et al. Influenza neuraminidase inhibitors possessing a novel hydrophobic interaction in the enzyme active site: design, synthesis, and structural analysis of carbocyclic sialic acid analogues with potent anti-influenza activity. Journal of the American Chemical Society. 1997; **119**(4): 681−90.

2. Sweeny DJ, Lynch G, Bidgood AM, Lew W, Wang KY, Cundy KC. Metabolism of the influenza neuraminidase inhibitor prodrug oseltamivir in the rat. Drug metabolism and disposition: the biological fate of chemicals. 2000; **28**(7): 737−41.

3. Abad-Zapatero C, Blasi D. Ligand efficiency indices (LEIs): more than a simple efficiency yardstick. Molecular informatics. 2011; **30**(2−3): 122−32.

4. Abad-Zapatero C, Perisic O, Wass J, Bento PA, Overington J, Al-Asikani B, Johnson M. Ligand efficiency indices for an effective mapping of chemicobiological space: the concept of an atlas-like representation. Drug discovery today. 2010; **15**(19−20): 804−11.

5. Abad-Zapatero C. Ligand efficiency indices for effective drug discovery. Expert opinion in drug discovery. 2007; **2**(4): 469−88.

6. Woltosz WS. If we designed airplanes like we design drugs. Journal of computer-aided molecular design. 2012; **26**(1): 159−63.

7. Huang N, Shoichet BK, Irwin JJ. Benchmarking sets for molecular docking. Journal of medicinal chemistry. 2006; **49**(23): 6789−801.

8. Mysinger MM, Carchia M, Irwin JJ, Shoichet BK. Directory of useful decoys, enhanced (DUD-E): better ligands and decoys for better benchmarking. Journal of medicinal chemistry. 2012; **55**(14): 6582−94.

9. Cortes-Cabrera A, Klett J, Dos Santos HG, Perona A, Gil-Redondo R, Francis SM, et al. CRDOCK: an ultrafast multipurpose protein−ligand docking tool. Journal of chemical information and modeling. 2012; **52**(8): 2300−9.

10. Manning G, Whyte DB, Martinez R, Hunter T, Sudarsanam S. The protein kinase complement of the human genome. Science. 2002; **298**(5600): 1912−34.

11. Knox C, Law V, Jewison T, Liu P, Ly S, Frolkis A, et al. DrugBank 3.0: a comprehensive resource for "omics" research on drugs. Nucleic acids research. 2011; **39**(Database issue): D1035−41.

12. Corey EJ, Czako B, Kurti L. Molecules and Medicine. Hoboken, NY: John Wiley & Sons; 2007.

13. Sussman JL, Harel M, Frolow F, Oefner C, Goldman A, Toker L, et al. Atomic structure of acetylcholinesterase from *Torpedo californica*: a prototypic acetylcholine-binding protein. Science. 1991; **253**(5022): 872−9.

14. Manglik A, Kruse AC, Kobilka TS, Thian FS, Mathiesen JM, Sunahara RK, et al. Crystal structure of the micro-opioid receptor bound to a morphinan antagonist. Nature. 2012; **485** (7398): 321−6.

15. Wu H, Wacker D, Mileni M, Katritch V, Han GW, Vardy E, et al. Structure of the human kappa-opioid receptor in complex with JDTic. Nature. 2012; **485**(7398): 327−32.

16. Granier S, Manglik A, Kruse AC, Kobilka TS, Thian FS, Weis WI, et al. Structure of the delta-opioid receptor bound to naltrindole. Nature. 2012; **485**(7398): 400−4.

17. Thompson AA, Liu W, Chun E, Katritch V, Wu H, Vardy E, et al. Structure of the nociceptin/orphanin FQ receptor in complex with a peptide mimetic. Nature. 2012; **485**(7398): 395−9.

18. Dortch-Carnes J, Potter DE. Bremazocine: a kappa-opioid agonist with potent analgesic and other pharmacologic properties. CNS drug reviews. 2005; **11**(2): 195−212.

19. Jhoti H, Leach AR. Structure-based drug discovery. Dordrecht: Springer; 2007.

20. Ward RA, Brassington C, Breeze AL, Caputo A, Critchlow S, Davies G, et al. Design and synthesis of novel lactate dehydrogenase A inhibitors by fragment-based lead generation. Journal of medicinal chemistry. 2012; **55**(7): 3285–306.

21. Adams MJ, Ford GC, Koekoek R, Lentz PJ, McPherson Jr. A, Rossmann MG, et al. Structure of lactate dehydrogenase at 2–8 Å resolution. Nature. 1970; **227**(5263): 1098–103.

22. Tanaka D, Tsuda Y, Shiyama T, Nishimura T, Chiyo N, Tominaga Y, et al. A practical use of ligand efficiency indices out of the fragment-based approach: ligand efficiency-guided lead identification of soluble epoxide hydrolase inhibitors. Journal of medicinal chemistry. 2011.

23. Blasi D, Arsequel G, Valencia G, Nieto J, Planas A, Pinto M, et al. Retrospective mapping of SAR data for TTR protein in chemicobiological space using ligand efficiency indices as a guide to drug discovery strategies. Molecular informatics. 2011; **30**(2–3): 161–7.

24. Mairal T, Nieto J, Pinto M, Almeida MR, Gales L, Ballesteros A, et al. Iodine atoms: a new molecular feature for the design of potent transthyretin fibrillogenesis inhibitors. PloS one. 2009; **4**(1): e4124.

25. Masciocchi J, Frau G, Fanton M, Sturlese M, Floris M, Pireddu L, et al. MMsINC: a large-scale chemoinformatics database. Nucleic acids research. 2009; **37**(Database issue): D284–90.

26. Wolber G, Langer T. LigandScout: 3-D pharmacophores derived from protein-bound ligands and their use as virtual screening filters. Journal of chemical information and modeling. 2005; **45**(1): 160–9.

27. MOE. Chemical Computing Group Inc., Montreal. Available from: <http://www.chemcomp.com>.

28. Gonzalez A, Quirante J, Nieto J, Almeida MR, Saraiva MJ, Planas A, et al. Isatin derivatives, a novel class of transthyretin fibrillogenesis inhibitors. Bioorganic and medicinal chemistry letters. 2009; **19**(17): 5270–3.

29. Adamski-Werner SL, Palaninathan SK, Sacchettini JC, Kelly JW. Diflunisal analogues stabilize the native state of transthyretin. Potent inhibition of amyloidogenesis. Journal of medicinal chemistry. 2004; **47**(2): 355–74.

THE ROAD AHEAD

Soon after the power of Harrison's chronometers was established to accurately determine the longitude at sea by Capt. Cook, the British Navy sent numerous expeditions all over the world to accurately map unchartered waters. Notable among them was the one aboard *H.M. S. Beagle* lead by Capt. Fitzroy to chart the waters of South America. For this mission, Capt. Fitzroy carried 22 Harrison chronometers. This expedition is now famous due to the biological insights of one of its passengers, Charles R. Darwin. The expedition of the *Beagle* resulted in a series of superb maps of the South America Coast line, including the Channel of the Beagle and the treacherous waters near the Magellan Strait.

The development of accurate maps of CBS based on LEIs (or other variables) will go a long way toward making drug discovery more predictable and not a "numbers game" but a "charted" journey. There will continue to be perils, rifts, sandbars, and other obstacles but holding onto our maps, expanding and revising them, and improving our navigational instruments we may be able to find the "treasure islands" that we have already hinted at in our limited traveling thus far. There will be new islands, peninsulas, and even continents rich with novel chemical matter related to promising biological targets. More importantly, we will be able to devise our optimal routes to travel from treasure to treasure in the most efficient and economical way. The dream of finding the "El Dorado" of chemical matter, where the majority of all medicinal drugs would be found might never materialize, as it has happened with other mythical dreams in space or time (i.e., the Fountain of Youth). Nonetheless, it can never be doubted that having an accurate representation of CBS and the appropriate instruments and milestones to navigate it are *sine qua non* conditions to venture into successful expeditions within these intricate waters.

The concepts introduced here and the AtlasCBS server should be just the beginning. Their limitations and means to overcome them have already been presented along the way as well as new directions and extensions of these ideas. I appeal to the community at large to explore these concepts and to continue to provide insights and impetus to make drug discovery more effective.

Algebraic Derivations

A.1 DERIVATION OF THE RELATIONSHIP BETWEEN nBEI AND NSEI AND RELATED VARIABLES IN THE CORRESPONDING EFFICIENCY PLANES

From the definitions in Table 2.1 for NSEI and nBEI, it can be shown that a linear relationship can be obtained relating nBEI and NSEI as follows. Using the definitions, one can eliminate the affinity variable ($\log_{10} K_i$) as follows.

$$NSEI = -\log_{10} K_i/NPOL \qquad (2.4)$$

$$nBEI = -\log_{10}[K_i/NHA] \qquad (2.6)$$

$$-\log_{10} K_i = NPOL \cdot NSEI \qquad (A.1)$$

$$-\log_{10} K_i = nBEI - \log_{10}(NHA) \qquad (A.2)$$

Substituting the value of $-\log_{10} K_i$ from Eq. (A.1) into Eq. (A.2) and rearranging terms:

$$nBEI = NPOL \cdot NSEI + \log_{10}(NHA) \qquad (A.3)$$

This equation describes a family of lines and represent the geometrical locus of all the target−ligand complexes in the chemical database, independent of the value of the variable chosen to describe the affinity. This is a linear relationship between nBEI and NSEI, where the slope (NPOL) and the intercept ($\log_{10}(NHA)$) depend *only* on the chemical properties of the ligand; the dependence on the target has been eliminated by eliminating the variable K_i (i.e., mathematical locus). The resulting distribution of the ligands in PDBbind in the Cartesian plane defined by nBEI vs. NSEI is presented in Fig. 2.2B.

Similarly, from the definitions in Eqs. (2.4) and (2.7) (Table 2.1), a linear equation can be also obtained relating mBEI and NSEI:

$$mBEI = -\log_{10}[K_i/MW] \qquad (2.7)$$

$$mBEI = NPOL \cdot NSEI + \log_{10}(MW) \qquad (A.4)$$

In Eq. (A.4), the intercept is $\log_{10}$(MW) that relates to the MW of the ligand. This representation is conceptually the same as Eq. (A.3) but allows a wider separation of the compounds on the vertical axis based on MW and permits a better way of monitoring the size of the ligands. An example of the distribution of the target–ligand pairs within WOMBAT, in the Cartesian plane defined by mBEI vs. NSEI, is presented in Fig. 3.2B (inset).

One more derivation should be considered to understand the appearance of a very unique efficiency plane. This is the plane obtained when the variables are SEI and mBEI (x,y) (Table 2.1). The use of this plane is very important to separate molecules in the angular coordinate related to their polar surface area (PSA). As indicated in the text, the separation by NPOL in the angular coordinate is not adequate in chemical structures with the same number of polar atoms but rather different 3-D structures, for example, polypeptides, natural products or even proteins with the same (or very similar) composition but different tertiary structures.

In these cases, the appearance of the recommended efficiency plane is obtained as follows. Using the definitions of mBEI and SEI from Eqs. (2.7) (above) and (2.3) (Table 2.1), renumbered here:

$$\text{mBEI} = -\log_{10}[(K_i/\text{MW})] \tag{A.5}$$
$$\text{SEI} = pK_i/(\text{PSA}/100) = -\log_{10} K_i/(\text{PSA}/100) \tag{A.6}$$

Expanding Eq. (A.5):

$$\text{mBEI} = -\log_{10} K_i + \log_{10}(\text{MW}) \tag{A.7}$$

and substituting $-\log_{10} K_i$ by its value in Eq. (A.6) we obtain:

$$\text{mBEI} = (\text{PSA}/100)\text{SEI} + \log_{10}(\text{MW}) \tag{A.8}$$

This equation permits a very fine separation of the molecular entities by 1/100 of the PSA value on the angular coordinate (slope). An example of this type of equation to separate peptides has been presented in Part II (Fig. 5.6B) using the analogous variables nBEI vs. SEI and in Part III (Fig. 8.13A) with mBEI vs. SEI to separate ligands for the opioid receptors.

Other efficiency planes that can be used to study the combined effect of size-efficiency vs. polarity-efficiency are: (i) BEI vs. NSEI;

(ii) NBEI vs. SEI; and (iii) nBEI vs. SEI. Following the examples above, it is quite simple to derive the slope and intercept of the corresponding lines of constant slope and also to understand the appearance of the planes. All of these combinations can be displayed and compared by going into the **Viewer** tab of the AtlasCBS tool allowing the analysis of any particular dataset in various efficiency planes.

APPENDIX B

Conversion Factors Among Various Ligand Efficiency Indices

B.1 BEI AND LE

From the definitions in Table 2.1 for BEI and LE, it is easy to obtain an approximate conversion factor relating the two efficiency indices, independent of the number of atoms or the molecular weight. The definitions are copied here and the equations are renumbered for convenience:

From Eq. (2.2):

$$BEI = -\log_{10} K_i/(MW/1000) \tag{B.1}$$

From Eq. (2.1):

$$LE = -RT \ln K_i/NHA \tag{B.2}$$

The scale factor between the two will be obtained by:

$$BEI/LE = [1000(-\log_{10} K_i)/MW]/[-RT \ln K_i/NHA] \tag{B.3}$$

Remembering that $\log_{10} K_i = \ln K_i \log_{10} e$, and substituting in the numerator:

$$BEI/LE = (-1000 \ln K_i \log_{10} e\ NHA)/(-MW\ RT \ln K_i) \tag{B.4}$$

The value of MW can be approximated as $MW \sim\ <MW_{atm}> NHA$. $<MW_{atm}>$ represents "mean molecular weight" per non-hydrogen atom on a ligand molecule, approximately 13.3 Da for a typical medicinal chemistry molecule[1]. This approximation yields:

$$BEI/LE \sim (-1000 \ln K_i \log_{10} e\ NHA)/(-13.3\ NHA\ RT \ln K_i) \tag{B.5}$$

Eliminating common factors in the numerator and denominator ($\ln K_i$, NHA, sign), we are left with an approximate conversion factor of:

$$BEI/LE \sim 1000 \log_{10} e/13.3 \cdot 0.594 \sim 54.8 \tag{B.6}$$

($R = 0.00198$ Kcal/degree mol) and $T = 300$.

B.2 LE AND NBEI

This conversion is particularly relevant because it relates two efficiency indices scaled by the atom count of the molecule and therefore show how all the different ways of defining ligand efficiency are interrelated except for scaling constants. Once again we copy the definitions of the efficiencies (Table 2.1) and renumber the equations:

$$LE = - RT \ln K_i / NHA \qquad (B.2)$$

$$NBEI = - \log_{10} K_i / NHA \qquad (B.7)$$

The conversion factor will be given by:

$$LE/NBEI = [-RT \ln K_i / NHA] / [-\log_{10} K_i / NHA] \qquad (B.8)$$

Cancelling out the common factors (NHA, $\ln K_i$) and rearranging the quotient, we are left with the following ratio of constants:

$$LE/NBEI = RT/\log_{10} e \approx 0.594/0.434 \approx 1.370 \qquad (B.9)$$

Using: $\log_{10} K_i = \ln K_i \log_{10} e$ ($R = 0.00198$ Kcal/degree mol   and $T = 300$).

This relationship does not depend on any approximation to estimate the MW of the compound and therefore, except for the limitation on the number of decimal places used for the constants, is very accurate. If BEI ~ 54.8 LE (Eq. (B.6)) and LE ~ 1.370 NBEI (Eq. (B.9)), it follows that BEI ~ 75.1 NBEI. It is left as an exercise to show that this is also true starting from the corresponding definitions of BEI and NBEI. Deriving other equivalence factors (e.g., SEI/NSEI and similar) could also aid in understanding the concepts discussed.

REFERENCE

1. Hopkins AL, Groom CR, Alex A. Ligand efficiency: a useful metric for lead selection. Drug discovery today. 2004; **9**(May): 430–1.

Printed and bound by CPI Group (UK) Ltd, Croydon, CR0 4YY

03/10/2024

01040420-0017